The *Coca-Cola* Trail

People and Places in the History of Coca-Cola

Larry Jorgensen

Modern History Press

Ann Arbor, MI

ISBN 978-0-692-84430-4

Published by
Modern History Press
5145 Pontiac Trail
Ann Arbor, MI 48105

info@ModernHistoryPress.com
Tollfree 888-761-6268
FAX 734-663-6861

Contact publisher for discount on bulk purchases for sales promotions, fund-raising, or educational use.

Foreword

The Coca-Cola Trail

So, what is a Coca-Cola Bottler? Well, from a legal standpoint, it's a franchisee of the Coca-Cola company, who has been granted the right to bottle and sell Coca-Cola in an exclusive territory. This franchise or contract has been amended and otherwise changed somewhat over the years to add many additional brands, but in the beginning, it was granted for the sole purpose of bottling and distributing Coca-Cola throughout the bottler's territory.

In the early years it was bottled one bottle at a time and placed in wooden cases and hauled to the customer in horse or mule drawn wagons. The bottling plants were crude and the task was difficult but by acquiring these franchises, these entrepreneurs had unknowingly won the lottery. They had settled on investing in a product that would change the world. A fabulous beverage with a delicious, unique taste profile that provided a refreshing boost to the drinker. Even in today's world, with thousands of brands of both non-alcoholic and alcoholic beverages, Coca-Cola remains at the top in providing that unique taste and special experience in a beverage. You combine this great product with brilliant advertising and marketing from The Coca-Cola Company and "voila" you have a brand and company and a bottling and distribution system that has been so successful that books upon books have been written about it for over 100 years. That is unique in itself.

But the history and legacy goes deeper. Year after year, decade after decade, one generation of bottlers led to the next generation and then the next and, in the case of my family and a number of other families in the business, we are well into our 5th generation in this business. From the first bottling of Coca-Cola in Vicksburg, Mississippi in 1894 until today, community support and leadership have been paramount. The relationships created between customers, consumers, suppliers, employees, management and ownership, and really everyone that has touched Coca-Cola, have evolved into a passion for this brand and its trademark like no other business in history.

Larry Jorgensen has captured the essence of the history and passion for this business through exhaustive interviews with the family members/owners and associates of these multi-generational enterprises. Many of these bottlers have museums of Coca-Cola memorabilia and historical information that most everyone will find to be a fun excursion to see and learn and enjoy. A history book and travel guide all rolled into one that will transport you back to another time and recall happy memories of days gone by.

March 22, 2017   RANDY MAYO
                                                        4th generation Coca-Cola Bottler

Introduction

The Coca-Cola Trail

During the first half of the 20th century, large and small towns throughout the United States received special recognition when they became the site for a Coca-Cola bottling plant. Additional pride came when the town's name was placed on the bottom of bottles from their plant.

By 1909 there were 400 community plants providing the exciting new Coca-Cola beverage for customers in their sales territories, which were granted to the plants by exclusive bottling agreements. The number of Coca-Cola plants increased to over 1200 by 1925, and most of them were local family owned businesses.

"The Coca-Cola Trail" traces the history of some of those plants, and of those pioneering entrepreneurs who saw what they believed to be an opportunity, and then took the risk to help create what would become one of the most recognized brand names in the world. Without their dedication and tireless efforts, it reasonable can be assumed the Coca-Cola empire may not have evolved as it did.

"The Coca-Cola Trail" is dedicated to those first bottlers, and in many cases, to the following generations of their families who today still continue to make the enjoyment of Coke products available to everyone.

As the demand for Coca-Cola grew, larger plants were needed in those communities, and often the new buildings became local landmarks. Everyone in town was familiar with the plant, and most had their own Coca-Cola memories. They often recall looking in a large glass window to watch the bottles being filled, or possibly enjoying a group tour of the plant. Also visible were the colorful trucks being loaded and heading out for daily deliveries. Others remember friends or relatives who worked in the plant. The special Coca-Cola promotional activities and community participation were common and appreciated.

However, by the 1960's the improvement in transportation and ease of distribution began to signal the end for small town bottlers. During the decades which followed the smaller plants were being closed, and bottling was being consolidated to larger more modern facilities. Some of the former plant buildings were utilized as warehouse and distribution facilities. Others would find a new purpose in the community, while some were simply torn down.

"The Coca-Cola Trail" visits some of the remaining Coca-Cola buildings to discover how they are being used today. The "Trail" also reveals the history of these facilities and the people who made them possible. The "Trail" begins in Vicksburg, Mississippi where Coca-Cola was first placed in a bottle to be sold. The "Trail" also visits museums and displays, famous old Coca-Cola signs, and other places of interest made possible by "the people of Coca-Cola".

It was, and still is, the large and small Coca-Cola companies across our nation which continue the Coca-Cola legend with community involvement, creative advertising, and unique marketing and promotions. "It's the real thing"!

The Coca-Cola Trail

Contents

Acknowledgements

The Coca-Cola Trail

This journey down "The Coca-Cola Trail" could not have been made without the support and assistance of many who wanted the story to be told!

Our appreciation is deep and sincere for those, in the chapters which follow, who took time to share their memories and history. But our appreciation is not limited to those dedicated bottlers, it also is for the researchers in historical organizations, museums and libraries, and for the Coca-Cola "fans" who also had stories to tell. The reader meets many of them along the"Trail".

However, special recognition and thanks is extended to a Coca-Cola bottler in Texas, who generously shared his unique knowledge of Coca-Cola history, as well as his pride in a way of life for generations of his family. Our appreciation always to Randy Mayo, a direct Biedenharn family descendent, who can trace his roots back to that first bottle of Coca-Cola in Vicksburg.

To all those who contributed time and effort, you helped smooth the bumps in the "Coca-Cola Trail".

Chapter 1
Vicksburg, MS

The Coca-Cola Trail starts in Vicksburg, Mississippi where in the summer of 1894 Joseph A. Biedenharn first put Coca-Cola in a bottle, and consequently started the world's first Coca-Cola bottling business.

Joseph was the son of a German emigrant, Herman Henry Biedenharn who arrived in Louisiana in 1852. After working in several Louisiana cities, Herman settled in Monroe, Louisiana where he met and married Joe's mother, Louisa Lundberg, who had emigrated from Denmark.

The Civil War took Herman Biedenharn to Vicksburg where his skill as a boot maker was pressed into service making and repairing boots and shoes for the Confederate soldiers. Louisa joined her husband in Vicksburg after the war, and Joseph Augustus Biedenharn, their first son, was born in 1866.

Joe's father continued his shoe and boot business, but the future Coca-Cola business would eventually become a reality because Herman also decided to open a confectionery store.

Herman's brother arrived from Germany to run the store, and at the age of 14 Joe dropped out of school to work full time with his uncle in the store. Joe became manager of the store when his uncle died in 1888. The first telephone in Vicksburg was installed in the Biedenharn business in 1889.

Joe added a soda fountain to the store, and in 1890 he and his father built a two story brick building at 218-220 Washington Street to house both the boot and shoe and the confectionery businesses. Joe's store had a long soda fountain where carbonated water drinks flavored with syrup were sold by the glass.

A difficult business situation resulted in Joe getting into the bottling business in 1891. He had received customer orders for 30 cases of soda water, and as he had in the past, went to the local bottling company to have his orders filled. However Joe's large order could not be done due to a heavy demand for soda created by the July 4th picnics being held that weekend.

That incident caused Joe to decide to begin bottling his own soda water. Second hand equipment was purchased in St. Louis and within 30 days he was in the business of bottling and distributing lemon, strawberry and sarsaparilla soda water.

A few years before the Vicksburg soda water incident, a pharmacist in Atlanta, John S. Pemberton had created the "secret formula" for Coca-Cola syrup. Pemberton had been injured in the civil war, and developed an addiction to morphine. He decided to invent a tonic to cure him of his habit. He created a "coca wine" made from coca leaves (the basis for cocaine) and kola nuts, a source of caffeine.

The syrup drink was offered for sale at drug store soda fountains, as Pemberton claimed it would cure several ailments. The first advertisement for Pemberton's Coca-Cola appeared in 1886.

In 1890 Coca-Cola traveling salesman Samuel Candler Dobbs persuaded Joe Biedenharn to buy a five gallon keg of syrup to serve as a new drink at his fountain. The salesman was the nephew of Asa Candler of Atlanta, who had recently purchased Coca-Cola from Pemberton for $2300. Samuel probably was Candler's first salesman.

The Biedenharn soda fountain served a 10 ounce and a 12 ounce glass of Coca-Cola, your choice for a nickel. In addition the bakery made jelly rolls and pound cake. A half-pound slice of cake also sold for a nickel.

The popularity of the fountain drink grew in Vicksburg and larger syrup orders were being placed. Asa Candler then paid a visit to Joe in 1891, and convinced him to become a wholesale distributor for the syrup. Owners of soda fountains up and down the Mississippi River and the surrounding area were solicited and became repeat customers for the syrup. Candler made at least two more visits to Vicksburg. The Biedenharns would give him a tour by delivery wagon to visit other soda fountains in the area. At that time Vicksburg had about a dozen soda fountains and a population of 20,000 people.

The fountain syrup sales continued to grow even after the bottling began. By 1897 the Coca-Cola Company owed the Biedenharn Company over $500 in commissions for the syrup sales. At that time Candler was offering his distributors one share of stock in his Coca-Cola Company for every $100 he owed them. This would have given Biedenharn Candy Company five shares . There were only 500 shares outstanding, which meant the Biedenharns would have owned one- percent of the company. However Joe needed cash to finance the growing needs of his company, and declined the stock offer. It was later estimated that in 1990 one-percent of the Coca-Cola Company was valued at over $300 million.

In the meantime, the Coca-Cola fountain drink had become such a success that Joe concluded it would be just as popular with people living outside of Vicksburg if they could have access to it. Thus the decision to bottle the drink was easy, as Joe already had purchased the necessary bottling equipment.

"The world's first Coca-Cola delivery man". Andrew Butler making a delivery in 1896 in Vicksburg.

The Coca-Cola syrup would arrive in wooden barrels. When empty the barrels were sawed in half, with one half being used to wash bottles. The other half would become a cooler where bottled Coca-Cola was placed in ice for sale.

Twenty seven year old Joe Biedenharn did not realize that summer of 1894, the far-reaching future impact of what he was doing when he poured the Coca-Cola syrup and shot the carbonated water into that first bottle. Joe had six younger brothers, most of them already involved at that time in the various activities of the Biedenharn family business. William was 21 years old, followed by Harry age 17, Lawrence 15, Herman Henry 13, Ollie 11, and Albert 7 years old. That simple decision to bottle Coca-Cola was the beginning of the Biedenharn Coca-Cola family legacy for those brothers and generations to follow.

Those first cases of Coca-Cola sold for 75 cents. Joe sent two of his first cases of bottled Coca-Cola to Asa Candler in Atlanta, who wrote back that it was fine. Joe later remembered that Candler never returned the empty bottles. As the bottling sales grew, the Biedenharns used metal cases which held four dozen bottles for shipping. Wooden cases which held two dozen bottles were used for local delivery. Metal cases which held six dozen bottles were used for a short time, but were too heavy. A full case of four dozen bottles weighed about 100 pounds. The bottles were packed upside down in the cases so the Coca-Cola would be mixed, and it was mixed again when the bottles were removed and turned right side up.

The plant also continued to bottle their other flavored sodas. It was said that sometimes to get a dealer interested in buying Coca-Cola, a few free samples might be included in a flavored water order.

Joe took time away from the Vicksburg business in 1903 when he went to New Orleans to establish and operate the Biedenharn Burnette Candy Company. While he was gone brother Harry took over management in Vicksburg. However the New Orleans venture was not a success and three years later Joe returned to begin traveling and selling wholesale the products of the Vicksburg business, which included Coca-Cola, soda waters, candy, bakery and some wholesale groceries.

It was during this time the Biedenharn company increased its capabilities with the purchase of the equipment and inventory of the Hill City Bottling Works of Vicksburg. The owner, W. E. Beck had decided to go out of business.

It is interesting to note that while Joe Biedenharn was the first to bottle Coca-Cola in 1894, he did not become a licensed Coca-Cola bottler until 1906. The rights to establish bottlers in the States had been sold in 1899 to Ben Thomas and Joseph Whitehead of Chattanooga. However, when Asa Candler of Atlanta granted those rights, he did not include portions of Mississippi where, he explained Coca-Cola already was being bottled. (*See Chattanooga Chapter*)

Whitehead acquired the rights to the southern half of the United States, after he and Thomas agreed to divide their business.

Whitehead died in 1906, and his close associate Charles Veazey Rainwater was designated secretary-treasurer of the parent bottler in Atlanta. Whitehead's business partner John Lupton selected Rainwater for the position because he realized Rainwater had worked extensively with Whitehead to establish the Atlanta plant and sign bottling territory agreements.

One of Rainwater's first projects was to convince Joe Biedenharn to sign the standard bottling agreements.

Rainwater, made several trips to Vicksburg to meet with Joe and discuss the issue of becoming a licensed bottler. Rainwater finally convinced Joe to sign a bottling contract by convincing him the contract would provide perpetuity for the business, along with advertising allowances and other privileges offered by the parent company.

Joe Biedenharn was first, but many around the world, would follow in his footsteps. Brother Albert once said, "it grew right from the grass roots, started by people with very little financial means."

The Coca-Cola museum in Vicksburg, located in the original soda fountain building, features bottling equipment like that used in 1894. The display in-

cludes one of the original bottles which is embossed with the identification "Biedenharn Candy Company, Vicksburg, Miss".

That first bottle was called the Hutchinson blob-top bottle because it was sealed with a rubber disk pushed into the bottle neck and held with a wire. The problem of cleaning the Hutchinson bottle was recalled by Albert Biedenharn during an interview before he died. He was seven years old when one of his jobs in the family business was to rinse the bottles. He explained it was not possible to sanitize the bottles because of the wire inside. You could not get a brush or anything inside the bottle, so if dirt was visible a metal shot would be added to the rinse water. Then he said, you would shake the bottle to cut the dirt loose.

It also was discovered that the rubber seal eventually affected the drink's flavor, especially during the warm summer months. In the early 1900's Joe switched to a straight sided bottle with a crown top. For the first few years the crowns (caps) were plain, and Coca-Cola probably was first to provide a printed crown. The new bottles also were embossed with Biedenharn Candy Company, and now included "Coca-Cola" in script style lettering at the bottom. Today's familiar Coca-Cola style bottle was approved for use by all bottlers in 1916.

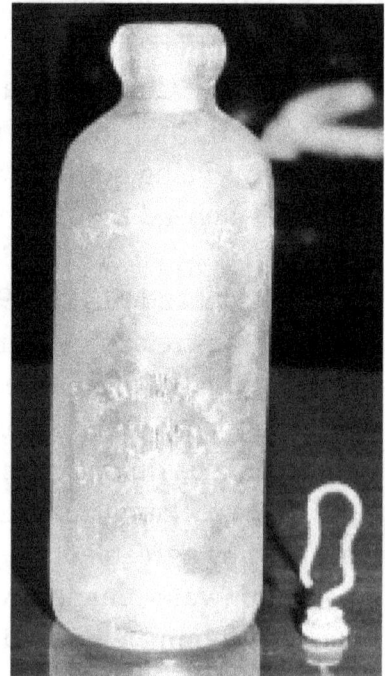

The Hutchinson Stoppered Bottle. The first to contain Coca-Cola.

The beginning of the end for the Hutchinson bottle came in 1906 when the FDA was established by the federal government to investigate food safety. The bottle came under attack because the agency said the wire allowed contaminant to enter the drink. Within a few years the Hutchinson bottles were no longer in use.

The Vicksburg museum also has displays of old vending machines, Coca-Cola advertising and other memorabilia. The restored candy store features a soda fountain where visitors can purchase Coca-Cola drinks, soda floats and souvenirs. In addition a display in the rear of the museum recreates the original 20' x 30' area where Coca-Cola was first bottled. The finished product was then carried through the store and loaded onto wagons in the front.

Continued growth of the business required a move to a larger facility. From 1902 to 1914 the bottling and candy manufacturing were done in a building on Grove Street, just around the corner from the original Washington Street location.

The next move came in 1914 when the operation was moved down the hill to the corner of Levee and South Streets. Will Biedenharn ran the plant when the move was made. Brother Albert helped and during the summer months actually ran the plant. The plant workers were expected to produce 15 cases an hour, or 150 cases during a ten hour day. The

This is a recent photograph of the building on Grove Street, which housed the bottling works and candy manufacturing from about 1902 to 1914.

plant was located close to the Mississippi River and sustained flooding in 1923, but continued at that site until 1938.

Biedenharn's last Vicksburg plant was built in 1938 and remained in operation at 2123 Washington Street until 1968. Brother Harry Biedenharn managed the Vicksburg bottling business from 1913 until his death in 1950.

The Last Coca-Cola plant in Vicksburg. Since being closed in 1966 has been used for local retail business. Coca-Cola cement cast identification remains.

Bottling at Vicksburg, like at several other area towns, was moved to Monroe, Louisiana after the Biedenharns opened a modern new plant there in 1966. This type of consolidation of plants was becoming more common in the business, as modern transportation and highways made it more efficient to transport than to maintain small community bottling plants.

The original site in Vicksburg was repurchased by the Biedenharn family in 1979. Old photos were used to restore the building which when completed, was donated to the Vicksburg Foundation for Historical Preservation, which now is responsible for the museum's operation.

In 1969 Coca-Cola Company officials in Atlanta predicted the future demise of the returnable glass bottle, saying by the 1980's it would be a "thing of the past". Consumers will want more beverages in cans and other convenient non-returnable containers.

Once again the Biedenharn family lead other Mississippi bottlers to create a way to meet the new challenges of canned beverages. Adding canning capabilities in each local bottling operation was not feasible.

Milton Biedenharn of Vicksburg was president of the Mississippi Council of Coca-Cola Bottlers. He supported the idea of creating a co-op of state bottlers, which would build and operate a single canning operation to provide all co-op members quality canned Coca-Cola products.

In 1971 the Gulf State Canners Corporation was formed by nine Mississippi bottlers, including Milton's brother Eric, also from Vicksburg. The brothers were sons of Harry Biedenharn.

Ultimately Clinton, Mississippi was selected as the site for the new plant, and Coca-Cola canning began there in 1973.

A series of colorful murals depicting Vicksburg's history, have been created on the city's river flood wall. The first mural, unveiled in 2000, depicts "the first bottling of Coca-Cola". It was painted by mural artist Robert Dafford of Lafayette, Louisiana, who is one of America's prominent mural artists. He has painted over 300 scenes throughout the United States, Canada and Europe. The Biedenharn family provided the funds to make the mural possible. A family spokesman praised the mural as accurately "depicting the family's interaction in the business".

Dafford also has painted Coca-Cola murals in New Orleans and Paducah, Ky.

To view the Vicksburg mural, it is just a short walk from the museum down to the river wall.

The Biedenharn brothers expanded their Vicksburg Coca-Cola enthusiasm to acquire or establish bottling plants in several southern and southwestern states, becoming one of the largest family Coca-Cola bottlers in the nation. But Vicksburg retains a rich Biedenharn heritage. In addition to the popular museum, more than 75 direct descendants still call the area home.

Another part of Biedenharn history was scheduled to become an attraction for Vicksburg visitors and residents in 2017. The Biedenharn family home, located just around the corner from the Bidenharn Candy Company Museum, had fallen into a state of disrepair and had been scheduled by the city for demolition, when it was discovered by a memorabilia collector and historian.

Dale Jennings envisioned a new life for the Biedenharn property when he purchased it in 2014. The purchase included the original family home of 3200 sq. ft. built in 1875 at 718 Grove Street. A 3600 sq. ft. addition was constructed in 1890. It was said the addition was built primarily for Henry Biedenharn's daughter Katie.

Also in 1890, Henry and son Joe constructed a new two-story building on Washington Street to house both the boot and shoe business and the candy business. Joe was 24 years old at the time.

Behind the original home, but still on the Biedenharn property, is a two story structure built in 1900, which also was included in Jenning's plans. Jennings said the "life expectancy" of the buildings when he bought them was in months not years.

"They were ready to come down." He explained what he is doing is best described as "building rehabilitation", which captures the character and features of the original structure, but repurposes it for a new commercial use.

Large beams and other wood, as well as metal salvaged during the rehabilitation work, has been re-used in all of the floors, doors and windows. Of special significance is the original stairway and railing, which were saved and completely restored. The stairway in the original home leads to a second floor private two bedroom condo with modern living quarters. Included on the lower level will be a commercial restaurant kitchen.

A new central connection joins the two adjacent structures to house the "Mississippi Barbecue Company" featuring a restaurant, bar and entertainment area. An event area was planned for the upper level of the addition structure, along with a patio overlooking the nearby Yazoo River.

Jennings is passionate about history and restoration, and said he always wanted to open a restaurant, so the Biedenharn project combines both his interests. Previously he has restored two homes, including his personal residence, but the downtown Vicksburg project is his largest venture.

Memorabilia collecting has resulted in Jennings acquiring a variety of rare Coca-Cola pieces, many to be featured in the restaurant. Also to be displayed will be old service station pumps and signs, as well as old barber chairs.

The historic significance of the building is commemorated at the entrance with a plaque paying tribute to the Biedenharns along with a dimensional casting of a Coca-Cola bottle. The polyurethane casting was obtained by Jennings from the cement logo sign on Biedenharn's last Vicksburg Coca-Cola plant on Washington Street.

The structure behind the original home was planned to become a tap house and include two multi-bedroom condo units. The large patio area between the two structures will become an outside entertainment area, and a food patio is planned to be adjacent to the bar-b-que restaurant.

In doing outside patio dirt work, several old Coca-Cola bottles were discovered, including an original Hutchinson stopper bottle. Jennings believed there would be more treasures to be found as work continued.

The tap room structure at the rear of the property has additional local history, being purchased in 1929 by a lady named Kate Skinner. It is said the building then was operated for three decades as a "working man's" house of prostitution.

Jennings fell in love with the Biedenharn buildings and said he didn't want them to go to waste, so he decided to create something for Vicksburg. His restaurant staff was ready to start work when the business opened in the summer of 2017.

Chapter 2
Monroe, LA

Monroe, Louisiana, about 75 miles west of Vicksburg, is the second stop on the Coca-Cola Trail, as it also was the second stop for Joe Biedenharn who had "given birth" to bottled Coca-Cola in his Vicksburg soda shop.

Monroe also is the site of the Biedenharn Coca-Cola Museum, which features displays and information about the legendary growth of the Biedenharn "Coca-Cola empire".

It began in 1912 when the Biedenharns' purchased a small plant, Ouachita Valley Bottling Works, located in Monroe. However, it was not Joe, but his brother Ollie who initiated the family's interest in expansion. Ollie was bookkeeper in Vicksburg when he attended a Coca-Cola convention in Atlanta in 1912, and at that meeting became enthusiastic about the Coca-Cola business.

Soon after returning to Vicksburg Ollie learned the plant in Monroe, owned by Joe Rynwick, was for sale. Ollie and his younger brother Albert visited Rynwick and the plant was purchased for the family for $20,000. Included in the purchase was a sales territory of nine area parishes.

Ollie Biedenharn shortly after purchase of plant, 1912

Ollie Biedenharn inside first plant in Monroe, 1912

Albert was first to move to Monroe to operate the new acquisition. However Joe later purchased sole ownership from the family and moved to Monroe in the summer of 1913. Albert returned to Vicksburg and remained active in the business there for six years until moving to Wichita Falls, Texas to manage another Coca-Cola plant which had been acquired by the family.

It was said Joe's interest in obtaining sole ownership in Monroe was due to his sons becoming of age to be involved in the business. He had three sons. Henry, age 21, became Vice President; and Malcolm, age 20, became Secretary-Treasurer in Monroe. Bernard, only six years old at the time, later would become involved as well and served as Chairman of the Companies' Board of Directors from 1961 to 1988.

A few months after acquiring the Monroe plant, Ollie was able to obtain a second plant for the family, with the purchase of Star Bottling Works in Shreveport for $40,000. Ollie moved to Shreveport in 1913 to become manager. One of the previous owners, H. S. Edwards stayed on as plant superintendent. Later Ollie's son, R. Zehntner Biedenharn would manage the Shreveport plant for 51 years, from 1935 until it was sold in 1986.

Shreveport plant - 1915-16

Rudolph "Jack" Zehntner on left, 1913. Notice straight sided bottles.

Ollie's strong belief in future success for Coca-Cola was clearly demonstrated at another Atlanta convention in 1914. At that gathering of bottlers he delivered a speech in which he said, "I have invested in Coca-Cola because I believe it is in its' infancy; that by energy properly exercised, the ordinary care that is given any business to perpetuate it, the proper safeguards taken for the public health, that our prospects for the future are brighter than our success has been in the past".

Baseball and other sports activities often provide a marketing opportunity for Coca-Cola bottlers. But in Shreveport in 1920 Ollie Biedenharn became totally involved with baseball as both a stadium and team owner. Biedenharn Stadium opened June 28, 1920 and was home to the Shreveport Sports baseball team until the stadium burned, May 4, 1932.

Biedenharn Stadium - Opening Day

Ollie Biedenharn also pioneered the concept of night baseball in July of 1930, when he hired General Electric to install a lighting system in his stadium. At a cost of $22,000 six towers were erected to support 148 lights and provide 200,000 candle power lighting. That first night baseball game in Louisiana put the Shreveport Sports in the forefront of a movement that revolutionized minor league baseball, and eventually had an impact on all levels of baseball in America.

Biedenharn saw night baseball as a solution to declining ballpark attendance, which was attributed to the discomfort of daytime heat and new area entertainment options, which included movies at a now air conditioned theater.

After the stadium burned, the Sports temporarily moved their Texas League home games to a stadium in Tyler, Texas. However Biedenharn Stadium was made playable again in about two weeks. The team franchise and the stadium ultimately were purchased by the Caddo Baseball Association.

Sports historians also tell of another stadium memory. It was May of 1921 when the Shreveport team played an exhibition game against the New York Yankees and their star slugger "Babe" Ruth. It was said that during pre-game batting practice the "Babe" hit nine balls out of the park, including one that crashed through the window of a distant passing street car. The Yankees claimed a lopsided victory as the slugger hit three home runs, including a grand slam. It was only a few days earlier, at a similar exhibition game in Monroe, where Babe failed to hit a single homer.

A family story has been told about a financial challenge which faced Joe in the early years at the Monroe plant. It was said the bank where he was doing business "went broke" and was closed. To survive Joe wrote all his customers explaining his situation and requesting whatever he was owed. Joe reported a "wonderful response" which allowed him to keep the Monroe plant operating.

The next family purchase came in 1914 when the Texarkana, Texas Coca-Cola Bottling Company was acquired. The original plant was located at 504 East Board Street. A new plant was built a few years later on Texas Avenue, with the current Texarkana plant opening in 1953 at 1930 New Boston Road.

First plant in Texarkana 504 East Board St., 1914

The Wichita Falls, Texas plant was purchased next in 1919. Albert managed that operation until 1931 when the family purchased a plant in San Antonio and Albert transferred there. Again in San Antonio Ollie had been instrumental in the transaction, as he had learned that L. W. Alexander wanted to sell. The San Antonio purchase included the Uvalde, Texas area which was being operated as a warehouse. The Biedenharns' built a plant there in 1941.

The Wichita Falls Plant

Albert remembered rough times when he first went to Wichita Falls. He described it as a rugged country with no paved roads, and the frequent rains would cause delivery trucks to become bogged down. Coca-Cola sold for $1.00 per case plus $1.00 for the deposit. Sales were slow during the cold winter months, he said "sometimes trucks would go out and they wouldn't even sell five cases".

Ultimately the Biedenharn family would operate twenty-six Coca-Cola plants in eight southern and western states.

The Temple Coca-Cola Bottling Company in Temple, Texas was added to the family "portfolio" in 1925. The plant was located at 117 S. First Street, but was relocated to the corner of Third Street and Nugent Avenue when a new plant was built in 1950. It marked the 25th anniversary of Biedenharn ownership in Temple.

Dixie Bottling Line - Temple, Texas, 1929

A westward expansion began in 1965 when bottling operations were purchased in Flagstaff and Winslow, Arizona. Next was a Coca-Cola plant purchased in 1980 in Cheyenne, Wyoming.

During those expansion years growth also continued in Monroe, with a larger plant being built in the 1920's, followed by a third larger and more modern facility opened in 1966 along U.S. Highway 165 at interstate highway 20. Production at smaller area plants was moved to Monroe and those plants were closed.

Ouachita Coca-Cola Bottle Co. - late 1920's

Ouachita Coca-Cola Bottling Co. of today. U.S. Highway 165 and Interstate 20, Monroe, Louisiana

1920 plant now ready for a "new life"

Joe Biedenharn died October 9, 1952. But with the foundation he had laid the Biedenharn Companies grew to become one of the largest Coca-Cola bottlers in the United States.

Joe's only grandson, Henry Alvin Jr. was General Manager in Monroe from 1950 to1976. The fourth generation, Henry Alvin III became manager in 1976 and the business was expanded into additional markets in Louisiana, Mississippi and Arkansas. Included were plants in Natchez and Jackson, Mississippi; Alexandria, Ruston and Tallulah, Louisiana; and Hope, Arkansas. Other areas which also had Biedenharn family plants were Vernon, Clarendon, Memphis, Kerrville and New Braunfels, Texas; Winfield, Kansas, McAlister, Oklahoma, and Leadville, Colorado.

In addition to their business success, the family often was involved in community civic, philanthropic, cultural, and religious activities.

One interesting experience in the family's business activity in Monroe is when they provided financial assistance for an up-start air service business, which ultimately became one of America's largest airlines. What is now Delta Airlines started in Monroe in 1925 as a crop dusting service. It became Delta Air Service in 1929 when it began flying passengers on a route from Dallas to Jackson, Mississippi with stops in Shreveport and Monroe.

The family received stock in the new airline company as financial assistance was provided on three separate occasions. Bernard Biedenharn, Joe's youngest son, became one of Delta's directors and accumulated more than one-half million shares of Delta common stock.

For many years Delta Airlines held its' annual meeting in a board room of a Monroe bank. Directors from throughout the United States would travel to Monroe for the meeting. It was reported that occasionally a Delta director who might own 50,000 or more shares of stock, would suggest that a growing company with the stature of Delta, should hold the annual meeting in a more fashionable location, such as New York or San Francisco. However, Bernard would refer to his more than 500,000 shares and respond to the suggestion by saying, "We'll see you next year in Monroe".

Monroe served as the headquarters for Delta until 1941, when it moved to Atlanta, ironically the home for another fast growing company, Coca-Cola.

Changes to the capitol gains tax rate in 1986 was one of the reasons the family decided to begin selling Coca-Cola plants it owned. The new plant in

Monroe was one of 14 facilities in three states, "The Biedenharn Bottling Group", which was sold to Coca-Cola Enterprises in Atlanta in 1995 for $313 million.

The ownership of the Monroe plant changed again in November 2015, when it was purchased by Coca-Cola United of Birmingham. Keith Biedenharn, the last Monroe link to the founding family, remains as sales manager in Monroe. Keith is the great-great-grandson of Joseph Biedenharn. Coca-Cola United also acquired the Shreveport facility, and has a new bottling plant in Baton Rouge, as well as other Louisiana operations in Lafayette, Alexandria, Lake Charles, New Orleans and New Iberia.

The Biedenharn family Coca-Cola tradition also continues in Texas, as Ollie's great grandson, Randy Mayo of San Antonio owns the franchises in Winfield, Kansas and McAlister, Oklahoma. His sons, Wes Mayo and Dolph Mayo, representing the fifth Biedenharn generation, are working at Sooner Coca-Cola Bottling in McAlister. Randy's daughter, Rainey Mayo Felts, is in the corporate office in San Antonio, working for both the McAlister and Winfield facilities.

The Biedenharn story is told, along with displays of memorabilia, at the Biedenharn Coca-Cola Museum, located along the Ouachita River at 2006 Riverside Drive in Monroe. Daily live presentations at the museum add colorful details to the story, and provide visitors an opportunity to ask questions and share their own Coca-Cola experiences. One popular museum attraction is a 1929 model T delivery truck, loaded with Coca-Cola cases and ready to make delivery.

A visit to the Coca-Cola museum also provides an opportunity to tour the nearby Biedenharn home, gardens and Bible museum. There the visitor will learn about the distinguished career of Joe's daughter Emy-Lou. Her love of music and vocal talent took her to Europe in the 1930's. She performed in numerous cities, including London, Hamburg, Berlin, Munich and Copenhagen. She received critical acclaim for her rich contralto voice.

The start of World War II forced Emy-Lou to return home to Monroe. There she created the formal garden which is part of the visitor's tour. Her inspiration for the garden was the many great gardens she had visited in Europe.

Another feature of the home tour is a visit to the Bible Museum, which Emy-Lou opened in 1971. The museum houses an outstanding collection of old Bibles and Biblical artifacts.

The Biedenharn home, Bible museum, garden and conservatory have been called one of the south's most celebrated cultural attractions. Public hours for the Coca-Cola Museum and Biedenharn home are 10:00 am to 5:00pm, Tuesday through Sunday. The last home tour starts at 3:30 pm.

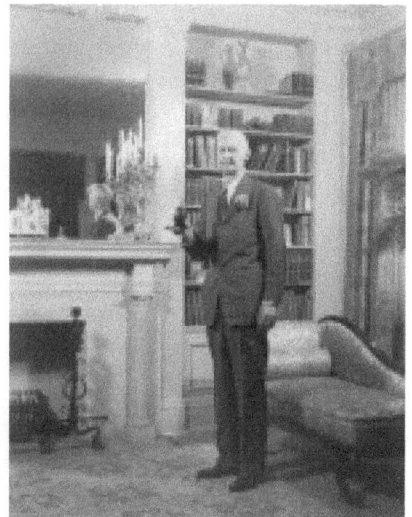

Joe Biedenharn - First bottler of Coca-Cola at his home in Monroe, Louisiana

Chapter 3
Chattanooga, TN

Almost everyone will recognize Atlanta as being the birthplace and official home of Coca-Cola. But others might claim Chattanooga has been the location for some of the most important events in the history of Coca-Cola. Some even say "it is the town Coca-Cola built", and the term "Coca-Cola money" is often mentioned in Chattanooga.

It all started with an idea by a soldier named Ben Thomas, who had been a Chattanooga lawyer and business man, before serving in the Spanish American war in Cuba. People in those days noticed Ben as someone in search of a big idea to make a fortune. It was that desire which caused him, while in Cuba, to notice the popularity of a carbonated soft drink called "Champagne Cola", which had a cold pineapple taste. Ben remembered the pleasing taste of Coca-Cola served at soda fountains back home, and became convinced of the potential of putting that taste in bottles. That idea matched his formula for success, "something inexpensive, which would be used up quickly and then repurchased".

When he returned to Chattanooga he first discussed his idea with a fellow boarder named Sam Erwin. The friend laughed at the idea, but agreed to help, as it happened that Coca-Cola President Asa Candler was Erwin's first cousin, and he was able to arrange a meeting for Thomas.

Thomas first traveled alone to Atlanta in 1898, but the trip was not a success as Candler did not seem interested in the bottling idea. Thomas decided he would need a business partner if he was to convince Candler of the sincerity of his proposal. He approached another boarding house friend, a fellow lawyer named Joseph Brown Whitehead. After considerable discussion, Whitehead agreed and the two men made the trip to Atlanta in the summer of 1899, to again meet with Candler. (some historians believe Candler's cousin, Sam Erwin accompanied Thomas and Whitehead on that trip.)

At first Candler was hesitant, fearing bottling would damage the reputation of his syrup company. It was Candler's opinion that bottling was a "back alley" business. He explained they already had bottled Coca-Cola briefly a year earlier, and he described the taste as "putrid". Candler added that he "had neither time, money nor brains" to attempt bottling. Thomas responded with a promise that he and Whitehead would guarantee the quality of the product, or if they failed to do so the agreement to bottle could be cancelled by Candler.

Thomas and Whitehead must have impressed Candler, as he finally suggested they prepare a contract for his consideration. A few days later the two men returned to his office with their proposal. The contract required the would be bottlers to use only Coca-Cola syrup; no substitutes. The soda fountain business remained Candler's and was excluded from the contract.

Also excluded was Mississippi where Joe Biedenharn had been bottling Coca-Cola for almost five years. Biedenharn had even sent two cases of his bottled product to Candler for approval. There was no reference made to Valdosta, Georgia where a small bottling company had been bottling Coca-Cola for two years. Candler may not have been aware of that situation, as the bottler was obtaining his syrup from a local soda fountain supply company. A "performance requirement" was included in the contract, and Candler then agreed and signed away, for a fee of $1.00, the bottling rights for Coca-Cola to the two men from Chattanooga. It is said Candler never bothered to collect the dollar. However he warned Thomas and Whitehead should their bottling idea become a failure, they should not come complaining back to him, as he explained he had very little confidence in the bottling business.

Benjamin Franklin Thomas and Joseph Brown Whitehead, obtained exclusive bottling rights for Coca-Cola.

Thomas and Whitehead returned to Chattanooga enthused about their new agreement, but also quickly realizing they now were faced with the giant task of creating a nationwide Coca-Cola bottling enterprise.

Thomas first made a quick trip to Shreveport to discuss the possibility of hiring an existing bottler. In fact, the very first few bottles under the new contract, were done in Shreveport as they experimented with the idea. However, Thomas was not satisfied with that option and returned to Chattanooga where he and Whitehead determined they would need about $7500 to establish their first bottling plant.

To help obtain the necessary funding Whitehead sold half of his share of the business to John T. Lupton of Chattanooga. Lupton also was an attorney, but had given up his practice after he married into the Patten family, the owners of The Chattanooga Medicine Company. They were making two popular proprietary medicines, so Lupton had experience which allowed him to see the potential of bottled Coca-Cola. Thomas retained his half ownership in the Coca-Cola bottling venture.

The first attempt to establish a bottling plant was brief and quickly aborted. The proposed site, an abandoned pool hall on Cowart Street, proved to be inadequate. What now is recognized as the first plant went into production in September 1899 in a small brick building at 17 Market St. Now designated Patten Parkway, it is the location of an historical marker at the original plant site.

That first plant was a small work place which created problems, and bottling was slow and difficult. The bottles were washed by hand, with metal shot often placed inside to help remove dirt during the cleaning process. Occasionally the pressure of the carbonation would cause bottles to explode, and workers often wore wire mesh face masks for protection. Spoilage was another problem, and it was reported that during hot weather the product only would last about ten days before turning rancid.

The first advertisement by the Chattanooga Coca-Cola Bottling Company was placed November 12, 1899 in the "Chattanooga Times". The message was simple and direct: "Drink a bottle of Coca-Cola, five cents at all stands, grocers and saloons."

Bottles of Coca-Cola were distributed by horse and mule drawn wagons. Often the driver would serve as a pitchman as well to attract attention to the new beverage. About three hundred gallons of Coca-Cola were bottled the first year of the plant's operation.

DRINK
A BOTTLE OF
Coca-Cola
5 CENTS
The most refreshing drink. Used summer and winter.
Sold at all stands, Grocer's and Saloons.

John Wilson Collection; Photos: Chattanooga Public Library

By April of 1900 the three partners agreed to divide their interests, in an attempt to better manage their vast enterprise. Thomas, who owned half of the company, placed a map on a desk and selected the territory for his share. The area included most of the states east of the Mississippi river and from Chattanooga north,

along with west coast states of California, Oregon and Washington. Lupton and Whitehead would concentrate on the deep south and remaining states west of the Mississippi. A second bottling plant was opened in Atlanta, and Whitehead moved there to operate it, while Lupton remained in Chattanooga, to ultimately create his own Coca-Cola bottling enterprise.

Whitehead established the first of several "parent companies" called Dixie Coca-Cola Bottling Company, and opened plants in nearly two dozen cities, from Savannah, Georgia to Meridian, Mississippi. The "parent" companies issued individual bottling contracts, (franchises) and took orders for syrup from the local plants. The syrup orders were forwarded to Candler, who shipped the syrup from Atlanta direct to the local plants. The parent companies were paid royalties on the syrup sales.

Whitehead died in 1906, and Lupton remained in Chattanooga and named Charles Veazey Rainwater as Secretary-Treasurer of the parent Atlanta operation. Rainwater had been a close associate to Whitehead, working to establish bottling territory agreements.

Asa Candler sold his interest in Coca-Cola to Bob Woodruff in 1919. Woodruff was not pleased with the royalties being paid to the parent companies, and began a plan to bring them under his control. In 1934 Woodruff bought out the original Whitehead-Lupton partnership. However, Lupton also had created, and maintained his own successful Coca-Cola bottling enterprise.

When Lupton granted bottling territories, often it was to relatives or friends he could trust. Many times Lupton would provide start up funding for the new bottlers. In return Lupton usually obtained an ownership for as much as half or even more in the new companies. These investments made Lupton the most influential person in the Coca-Cola bottling industry. He became wealthy because of profits received from the bottling companies, as well as from syrup royalties received as a parent company.

With the ready cash from his Coca-Cola business Lupton was able to invest in a variety of business and community projects in Chattanooga. Between 1900 and his death in 1933, Lupton greatly expanded his business empire and his family's fortune.

Lupton's son, Thomas Cartter Lupton, consolidated and managed the businesses until his death in 1977. Then Cartter's son, Jack Lupton created the JTL

Corporation to again consolidate and manage the family businesses, including all those relating to Coca-Cola.

However in 1986 the eighty-seven year Lupton family link to Coca-Cola came to an end. Jack Lupton sold all the family's bottling firms to Coca-Cola in Atlanta for $1.4 billion. At that time Lupton was the largest Coca-Cola bottler in the world, with an empire which included large plants in Dallas, Houston and Denver.

Meanwhile the other partner in the original bottling contract, Ben Thomas also soon realized there was more potential in selling Coca-Cola franchise territories than there was in the actual bottling business. So in 1902 Thomas sold his original Chattanooga plant to James F. Johnston, a farmer from Cleveland, Tennessee. Johnston, a friend of Thomas, also was a stockholder in the bottling company. At the time of the sale the plant's annual production had increased to several thousand gallons.

As Thomas aged he became concerned about who would continue to head his "parent" company, as he had no children. In 1904 his nephew George Thomas Hunter of Marysville, Kentucky began part time work at Johnston's plant while attending Baylor school in Chattanooga. Two years later Hunter was named Secretary of the company. Thomas remained committed to his pledge to Candler and continued to devote his efforts to the expansion of Coca-Cola bottling. It was reported that by 1909 there were 400 Coca-Cola plants operating in the United States, just ten years after the contract to bottle had been granted.

Johnston operated the Chattanooga bottling business until 1924, when he sold it to Crawford Johnson of Birmingham. Johnson already was in the Coca-Cola business as he had left his job as county clerk in 1902 to establish Birmingham's first plant. The Johnson family and its subsequent corporation continues to

New bottlewasher being delivered to Coca Cola Plant

operate Chattanooga as an important part of their bottling empire which has grown to become the third largest independent Coca-Cola bottler in the United States.

Crawford T. Johnson Jr. became President in 1942, followed by Crawford T. Johnson III in 1963. The Johnson family continued to acquire territories throughout the southeast, and in 1974 all the companies were consolidated into one company, Coca-Cola United. The company remains headquartered in Birmingham, serving seven southeastern states with 35 sales and distribution centers and three production facilities, like the one in Chattanooga. Claude B. Nielsen, son-in-law of Crawford Johnson III, is President and CEO of United.

The Chattanooga plant moved to its current location at 4010 Amnicola Highway in 1970 and in 1999 celebrated its 100th anniversary as the world's first licensed Coca-Cola bottler.

The community participated in the centennial observation with a variety of events and activities. Grade school children assembled a 14 by 40 foot birthday card in the shape of a Coca-Cola bottle. A sculpture designated "Second Century Coca-Cola Bottler" was created by renowned Atlanta artist Sergi Dolfi, who also worked for Coca-Cola Corporate for over 40 years. He was with Coca-Cola export, working in Milan, Rome and London from 1947 until 1973, when he

returned to Atlanta. He then was Coca-Cola Director of Corporate Guest Affairs until he retired in 1987. Dolfi was an active sculptor for more than 50 years.

Commemorative Coca-Cola packaging was created for the event, and downtown Chattanooga streets were the scene for a parade by Coca-Cola delivery men dressed in 1930's uniforms and driving yellow antique delivery trucks.

In 2014 Chattanooga Coca-Cola officials announced plans to build a new state-of-the-art distribution and sales facility. A ground breaking ceremony was held July 21st, a date which marked the 115th anniversary of Coca-Cola in Chattanooga.

The new 305,000 square foot distribution center was built at a cost of $67 million, and was officially dedicated March 1, 2016. The new building, when combined with the production facility on Amnicola Highway, provides Coca-Cola

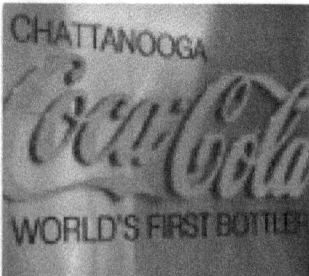

United with more than 600,000 square feet of space in the city where it all got started. The Chattanooga operation serves more than 6400 retail and restaurant customers in twelve counties in Tennessee, Georgia, and Alabama.

Chattanooga Glass Company

The city of Chattanooga may be recognized first as the birthplace of the Coca-Cola bottling industry. However, that historic development was equally significant for several other businesses which evolved to meet the growing needs of those new Coca-Cola bottlers.

One such company was Chattanooga Glass Company which began operation in 1901 to manufacture beer bottles, but ultimately became one of the nation's major producers of Coca-Cola bottles.

Charles Reif founded the glass bottle company to produce beer bottles for his Chattanooga Brewing Company. The bottle company grew quickly as he soon was able to double his production by using a new high grade sand which had been discovered at nearby Lookout Mountain.

In 1916 the company began producing the now famous Coca-Cola bottle which had been designed by Root Glass Company in Terre Haute, Indiana. The Chattanooga company made a first attempt at expansion a year later with the purchase of a glass company in Tallapoosa, Georgia. That plant closed three years later, but Chattanooga Glass would make other acquisitions in other cities in future years.

With the advent of prohibition in the 1920's, Charles Reif began to produce a near beer and other products, and changed his brewing company name to Purity Extract and Tonic. He also phased out of his involvement in the glass company, but remained a good customer purchasing bottles for his new products.

The Lupton name became connected to Chattanooga Glass in 1925, when John Lupton's nephew, J. Frank Harrison was one of three businessmen to purchase the company. Harrison became president and the company focused on providing bottles for Coca-Cola plants throughout the country.

Harrison died in 1933, and Clarence Renshaw Avery, who had been secretary, became President. The business continued to achieve record growth, and in 1953 J. Frank Harrison Jr. became President. Harrison also was board chairman for Coca-Cola Bottling Consolidated. At its peak production in the early 1960's the glass plant was making 610,000 bottles per day, operating with 600 workers on each of three eight-hour daily shifts, seven days per week. The company's annual payroll at that time was over $3.5 million. It had customers in forty states.

Harrison was president in 1960 when Chattanooga Glass was purchased by the Dorsey Corporation of New York. Harrison predicted continued growth for the company, and a second plant was built in Corsicana, Texas to produce over 15 million bottles per year. Ultimately the company would operate five plants throughout the U.S.

It also was in the 1960's when Coca-Cola corporate officials advised bottlers of pending changes in beverage packaging. They saw the consumer becoming more attracted to the convenience of "throw-away" bottles and cans.

Herbert L. Oakes, president of Chattanooga Glass, in 1965 discussed the issue of non-returnable bottles and cans. In noting the cost of a returnable bottle was twice as much as a non-returnable, Oakes emphasized the actual value of the returnable, as it would be re-used 25 to 30 times. However, he added that Chattanooga Glass was not ruling out the possibility of producing non-returnable bottles should the market demand them.

On the subject of canned soft drinks, Oakes emphasized the large cost necessary for individual bottlers to convert their equipment for cans.

The company's director of sales and advertising, Lupton Avery also saw the need to promote the advantages of glass bottles.

"It has no chemical effect on the food or the taste", he said during a 1965 speech to a local Rotary club. He noted that glass could be sealed tight preventing contamination, it can be formed into almost any shape, and people are able to see what they eat or drink. Lupton, the son of C. R. Avery, was employed at Chattanooga Glass from 1940 to 1985. One of Avery's most unique projects was to coordinate the company's observance of its 75th anniversary. The celebration had special significance as it occurred in 1976, which was coordinated with the U.S. Bicentennial observance.

The owner of Chattanooga Glass, the Dorsey Company, closely aware of rapid changes in beverage packaging, by 1978 also was operating the Sewell Plastic Division, which was quickly making inroads into the soft drink market with a two litre PET container.

Dorsey Company officials stated glass companies will have to find new markets and "even struggle to survive over the coming years". They sited the new 7-ounce beer bottle as one of those new markets.

The Dorsey company "threw in the towel" in 1982 and sold Chattanooga Glass to members of the firm's management team along with outside investors, for $40 million.

The plant changed hands again three years later when it was acquire by Diamond Container of Philadelphia. However, confronted with the beverage industry's growing demand for cans, the Chattanooga Glass plant was closed for good in 1988. Business assets were acquired by Anchor Glass, and finally sold again to a company in Mexico.

Several Coca-Cola bottlers remember it becoming extremely difficult to obtain bottles from Mexico.

CHRONOLOGY OF THE COCA-COLA BOTTLE (1984 - 1970)

1894 1899   1900 - 1916   1915   1923   1937   1957   1961   1970

Chattanooga Labeling Company

The spirit of the Chattanooga Glass Company lives on today with businessman Marvin Smith, who was an employee of the glass company for ten years. Smith worked in the bottle decorating division at company plants in Chattanooga, Corsicana, and Mt. Vernon, Ohio.

Smith started Chattanooga Labeling Company in 1991 when he saw an opportunity to do decorated bottles for Coca-Cola in Atlanta. Included was a special Coca-Cola Christmas bottle in 1995. For the first five years his 30,000 square foot plant concentrated much of its production on Coca-Cola bottles. However other companies soon became customers, and decorated items such as spirits bottles and jelly jars were being produced.

Growth continued for the labeling company, and in 2007 Smith returned to the original Alton Park site of Chattanooga Glass, where he built and operated a new plant for over five years.

Today Chattanooga Labeling is recognized for providing expert glass decorating services to the largest brands in wine, spirits, alcoholic and non-alcoholic beverages; as well as food companies and a wide variety of bottling customers.

With a new plant in Chattanooga and a second plant in Galeton, Pennsylvania, the company decorates over 100 million pieces of glass per year. Smith and his 75 employees in 2016 observed the 25th year in business for Chattanooga Labeling Systems; another area business with a connection to the Chattanooga Coca-Cola story.

Chattanooga Box and Lumber

For many years Coca-Cola bottles were placed into wooden cases, which today often serve as collector's items or decorative pieces. In 1917 a Chattanooga company discovered making those cases would become a way to be part of the rapidly growing bottling industry.

The company became one of the nation's largest producers of Coca-Cola cases. However during those start-up years it also survived by making World War I ammunition boxes for the government.

The company was known as Chattanooga Manufacturing in 1930 when it experienced financial difficulties, and was operated in receivership of a Chattanooga bank. In 1933 business partners Don Overmeyer and George Forman purchased the company, later to be named Chattanooga Box and Lumber Company. Overmeyer actually had joined the company in a sales position a few years earlier. Forman died in 1950, and Overmeyer acquired the other half of the business.

War again brought challenges and changes to the company. Sugar rationing during World War II reduced production and the need for wooden cases at Coca-Cola bottlers. However, once again the plant was called upon by the government to manufacture ammunition boxes. During its history, the plant also made cases for brewers such as Schlitz and others, as well as shipping crates for appliance manufacturers.

Overmeyer concentrated much of his efforts on obtaining box contracts with Coca-Cola bottlers throughout the country. Often he would accompany C. R. Avery of Chattanooga Glass Company on lengthy sales trips selling "boxes and bottles". Many times the wooden cases ordered by the bottlers, first would be delivered to the glass company to be filled with empty bottles before being shipped from Chattanooga.

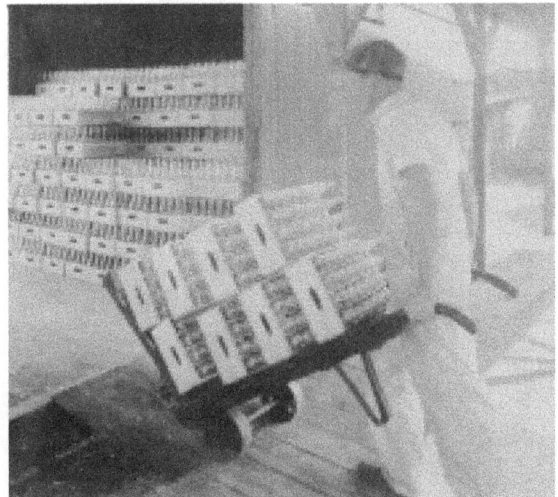

In 1960 a late night fire did extensive damage to the plant and destroyed several thousand dollars worth of lumber. Almost every available fire company and 60 firemen battled the blaze with flames soaring up to 300 feet in the air. It would be several weeks before production was resumed.

During the company's 50th anniversary in 1966, it was announced that the production of the 50,000,000 wooden beverage case had been achieved. At that time it was reported over 100,000 cases were being made each week for Coca-Cola bottlers.

Overmeyer remained as President in 1969 when Chattanooga Box and Lumber was sold to Temple   Industries of Diboll, Texas. At that time Temple also was making Coca-Cola cases, and was anxious to expand that business. The   Chattanooga plant had 155 employees and a spokesman said the plant would continue the production of its major product, beverage cases; and at the same time expand into the production of other wooden products.

However, it was a time when beverage packaging was changing more to the consumer preference of cans, and by 1978 employment at Chattanooga had declined to 95 workers. In September of that year Temple officials announced the closing of the plant.

Temple Vice President Harold Maxwell, said production in Chattanooga no longer was feasible, saying the plant "did not lend itself to extensive modernization", which was needed to meet environmental and safety standards. Temple continued production at plants in Texas and Savannah. The company was owned by Arthur Temple, who also had extensive timberlands and sawmills in east Texas.

The production of Coca-Cola cases in Chattanooga had an impact in the Linden, Alabama area as well. In the 1950's Don Overmeyer had found it becoming difficult to obtain enough lumber for his wooden box business. He decided to build a sawmill at Menlo, described as the "forest breadbasket" near Linden. There he manufactured cottonwood lumber used to construct the Coca-Cola cases, until a fire destroyed his mill in the 1960's, shortly before the Chattanooga plant was sold.

Overmeyer's son, Donald Jr. saw another opportunity, purchased the property and built a new sawmill. Today, Linden Lumber Company, under the third generation leadership of Hugh Overmeyer, markets its southern hardwood lumber internationally.

Cavalier Company

Coca-Cola fans who collect, or are interested in memorabilia, are familiar with the name Cavalier as one of the first, and for many years, the only manufacturer of Coca-Cola vending machines.

The company actually started in Chattanooga in 1865 as a sawmill, The Tennessee Furniture Company. It was an offshoot of the mill, created to sell flawed pieces of wood which might not have sold otherwise.

Gaston Raoul purchased the company in 1915, and the journey to become a major Coca-Cola supplier began when Raoul also purchased the Odorless Refrigeration Company. The company grew rapidly to become the largest manufacturer of ice coolers in the country. The name "Cavalier" was selected in 1923 for the company's line of soda coolers and cedar hope chests. During the depression of the 1930's the Cavalier Division began a long standing relationship with Coca-Cola. Under the name "Cavalier" they made soda machines, coolers, and picnic chests to specifications from Coca-Cola. The company's name officially changed to Cavalier Corporation in 1938.

John Wilson Collection
Photos: Chattanooga Public Library

William Gaston Raoul joined his father in the operation of the company in 1943, and Cavalier made products exclusively for Coca-Cola through the 1960's. The company was the first to make a vending machine which could dispense either cans or bottles.

The company proclaimed in 1958 that more than half a million coolers had been sold since 1935. Meanwhile the Tennessee Furniture Company continued to make furniture until 1960.

Foreseeing what may be rough times ahead, William Raoul held a Cavalier national sales conference in Chattanooga in May 1960. He noted it was the first time in more than thirty years all the sales representatives had been

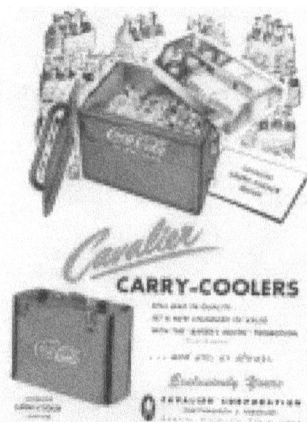

brought together. During the three day meeting Raoul reviewed the company's history and focused on future plans for Cavalier.

The Cavalier Corporation was acquired by the Seeburg Corporation in the 1960's. The name Seeburg is best remembered as a manufacturer of jukeboxes. The company actually got started making electric pianos in Chicago in the 1930's. In addition to the Cavalier purchase, Seeburg bought other companies in the 1960's, including those which made pinball and game machines.

Seeburg was purchased by H.N. White Company in 1966, and the music division's named was changed to King Musical Instruments, and it continued to make jukeboxes well into the seventies. However, the company apparently was unable to keep pace with the technical changes caused by the growing CD market. King was finally sold and closed.

Meanwhile the Cavalier Division filed for bankruptcy in 1987, and was spun off as an employee owned company. Cavalier was closed in August 2000.

Chapter 4
Valdosta, GA

Joseph Biedenharn is recognized as being the first person to bottle Coca-Cola in 1894 in his candy store and soda fountain in Vicksburg, Mississippi. Three years later a bottler in Valdosta, Georgia apparently became the second person to put Coca-Cola in a bottle, and again it was before the actual rights to bottle had been granted by Asa Candler in Atlanta.

It all started when H. H. (Hardy) Holmes opened a bottling plant in Valdosta in 1894. His brother Joseph became a partner in the business two years later. Hardy Holmes also had a son named Joe, who moved to Douglas, Arizona where later he would establish his own Coca-Cola plant.

Joseph Holmes

Soon after brother Joseph became a partner in the Valdosta plant, Hardy sold his share to Eugene Barber of Helena, Georgia. Barber and Joseph were married to sisters, and Barber moved his family to Valdosta.

E. R. Barber in office

In the late 1890's the plant began to bottle Coca-Cola, with syrup being purchased from a local wholesale fountain and drug supply company. The bottle used was the old Hutchinson style bottle with the metal stopper and rubber washer.

The majority of the bottling company's business was out of town orders, with the products being shipped in crates which held 72 bottles. To attract interest in the new Coca-Cola drink, six bottles would be included in every shipment.

Soon an odor problem was discovered in the Coca-Cola placed in the Hutchinson bottle, which caused Barber to search for a solution. The problem was solved when Valdosta became the first Georgia bottler to use the new crown cork and seal bottle. Barber purchased two of the foot powered machines and used them to meet the growing demand for bottled Coca-Cola.

Other drinks being bottled by the company included the popular Red Race ginger ale, Wards Orange Crush and Cherry Blossoms, Delaware Punch, NuGrape, a drink similar to 7-up called 2-way, strawberry and peach sodas, and vanilla and chocolate sodas. The plant became a franchised Coca-Cola bottler in 1903.

Holmes and Barber were the exclusive manufacturers of the Red Race ginger ale, which became a popular soft drink and was sold throughout a 100 mile radius of Valdosta.

It also is interesting to note that an Atlanta bottler started producing a product called Red Rock ginger ale in 1865.

THE WOMEN,

THE CHILDREN,

THE WHOLE FAMILY

can make life happy
by using carbonated
drinks from the

"It is a pleasure to take off the Crown."

Valdosta Bottling Works,
HOLMES & BARBER, Proprietors.

Have established more successful plants of that kind than any other firm in southern Georgia. They make a class of goods entirely their own, and they HAVE NO COMPETITORS. They expect to open a new bottling works in an adjoining town soon. They are originators of all high grade carbonated drinks, and are so recognized by the public. They were the first people in Georgia to adopt the famous "Crown" system of bottling, and the demand has grown so great that they will install a new machine at once to put up 24 bottles a minute.

Send Them Your Orders.

By 1906 Holmes and Barber were able to produce 1400 dozen bottles per day, and had twenty employees.

Early Coca-Cola promotion parade float and wheel barrow.

Barber remained active in the bottling business until 1923, when he sold out to his partner, J.F. Holmes. When Holmes died in 1936 he left the business to his three sons and three daughters. His will also contained a provision which required the likeness of a Coca-Cola bottle to be carved into his gravestone. It is said to be the only gravestone in the world to include a Coca-Cola bottle.

The husband of one of Holmes' daughters, William H. Warwick became vice president and manager, and the plant grew to become one of the largest in south Georgia, providing regular route service to five counties. Warwick also saw construction of the company's final plant in the early 1950's. A grand opening celebration attracted more than 15,000 Coca-Cola fans.

By 1965 many of the Holmes siblings had died, and Warwick was faced with the task of negotiating the sale of the business to the Frank Barron family of Rome, Georgia. Barron was a pioneer Coca-Cola bottler, and the family had plants in six other Georgia cities, including the first one in Rome, established in 1901.

The Valdosta plant was sold for just over $1 million, and remained a Barron family operation until 1986, when the Barrons sold all their plants and territory to Coca-Cola Enterprises in Atlanta for $84 million.

Bottling in Valdosta actually had stopped in 1981, but the facility continued as a distribution center; and in 1997 when the business observed its 100th anniversary, it was noted that over 100 Coca-Cola products were being distributed.

The Valdosta facility was closed and put up for sale in 2011. It was later acquired by East & West Investments of Valdosta, and in 2016 a search was underway for a new tenant. The Valdosta territory became part of Coca-Cola United in Birmingham in 2015.

Coca-Cola bottling in Valdosta is recorded as having taken place at four different plant locations through the years. The first plant at 105 N. Ashley was the home for Coca-Cola bottling for over 15 years.

One of the more interesting locations is the second site, which was in a building built in 1896, and which was still in use in 2016. The three story structure at 111 N. Ashley housed the Holmes and Barber plant from 1919 to 1925. Prior to that the building was owned by A. S. Pendleton Wholesale Grocery, and may still have owned it when the bottling business was

there, but could have been rented to Holmes and Barber. The structure, listed on the National Register of Historic Places, was purchased in 2016 by the Coleman Talley Law Firm, and was being remodeled to become the firm's new location. The new owners were planning to maintain some Coca-Cola identification

in their new offices. A street located alongside the building was listed in an old city directory as "Coca-Cola Alley".

Coca-Cola bottling relocated in 1925 to 115 Savannah, where a growing fleet of delivery trucks were accommodated. The final move came in 1952 when a modern

new facility was constructed at 1409 S. Ashley. When the Barron family bought the plant in 1965 they also maintained the old Savannah location as a warehouse and sign shop.

The Lowndes County Historical Society and Museum has a display of Coca-Cola memorabilia, as well as extensive documented files of Valdosta Coca-Cola; and one of the historians, Harry Evans will share personal memories as his father was a local Coca-Cola employee.

Valdosta Coca-Cola in 2016 at 2296 Hwy. 84

Photos and research special credit and appreciation to Lowndes Historical Society and Museum

Chapter 5
Root - The Bottle, The Bottler

Two museums which are over 900 miles apart, together reveal the fascinating history of the glass company responsible for the distinctive design of the Coca-Cola bottle; and the family which owned the glass company and ultimately became one of the nation's largest Coca-Cola bottlers.

It started in Terre Haute, Indiana where the Vigo County Historical Museum features a major collection commemorating the birth place of the Coca-Cola bottle, along with memorabilia, photos and information from the Root family.

It was Chapman J. Root, a Pennsylvania native, who opened Root Glassworks in Terre Haute in 1901. Three years later his company was manufacturing bottles for the Coca-Cola Company in Atlanta, and by 1912 over 800 people were employed at the glassworks. Root Glass owned a 120-acre silica sand plant near Terre Haute, which supplied raw material for the bottle production. In 1913 a tornado hit Terre Haute causing loss of life and damage to much of the city, including extensive damage to the Root plant. However, Chapman rebuilt in time to take advantage of a 1915 Coca-Cola challenge.

Chapman J. Root and Alex Samuelson

Courtesy of the Archives, The Coca-Cola Company

In 1915 Coca-Cola recognized the need for one distinctive bottle which would be used by all bottlers, and a nationwide competition was initiated to determine that bottle's design. Production was slow at Root Glass due to summer shut down, so Chapman took the Coca-Cola challenge and turned the project over to his management team: Alexander Samuelson, Plant Superintendent; T. Clyde Edwards, Auditor; Roy Hunt, Secretary; Earl R. Dean, Mold Shop Supervisor; and William R. Root, Chapman's son.

In researching an idea for the design Edwards found a drawing of a cocoa bean pod and he and Dean presented Samuelson a bottle sketch which featured a bulging middle, vertical grooves and tapered ends. After producing a few bottles they realized the large center bulge created a problem for the bottling equipment, and the team used the summer to redesign their idea.

Finally at a bottler's convention in Atlanta in 1915, the Root bottle was judged best because of potential customer recognition, ease of handling and production. Alexander Samuelson's name was on the bottle patent when issued November 16, 1915.

The greenish tint in the glass in the original bottle was due to the local silica sand. The color became traditional for Coca-Cola bottles.

The Root Glass Company received an exclusive contract to produce the bottle, and received a royalty of 5 cents for every 144 (gross) bottles produced. Demand for the new bottle grew rapidly, and at the peak of production in Terre Haute the company was producing 15-million Coca-Cola bottles per year. Root also licensed other glass companies to produce the patented bottle. Coca-Cola finally acquired the rights to the bottle in 1937.

Earl R. Dean stands beside the Company machine that produced the bottle

Coca-Cola was first bottled in Terre Haute in 1904 when the territory franchise was purchased by two businessmen, Edgar Coffee and Elmer Souderi. They began the business with foot powered bottling machines and horse drawn delivery Wagons. That first little plant was located at 924 Lafayette Street, and through the years and the many plant changes, 924 Lafayette Street has remained the address for Coca-Cola in Terre Haute.

Chapman Root bought the plant, the first in what would become the Associated Coca-Cola bottling empire, one of the largest independent bottlers in the nation. Root sold his glassworks in 1932 to the Owens-Illinois Glass Com-

Photos and Research: Vigo County Public Library

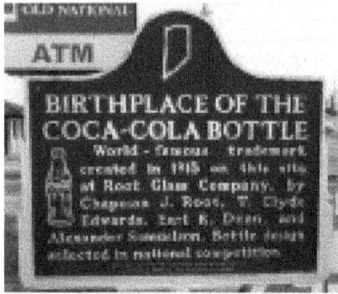

pany. The business experienced three more changes before finally being closed in 1984. The buildings were razed in the 1990's and a state plaque now marks the location of the birth of the Coca-Cola bottle.

Chapman Root's son, William R. Root, was killed in a plane crash in 1932 near Farmersburg, Indiana. Willliam's friend Paul Cox, who was piloting the plane, also died in the crash. Cox was a World War I hero and had spent 15 months overseas instructing other aviators. The old Terre Haute municipal airport, Dresser Field, was renamed Paul Cox Field in 1933. Chapman died in 1945, and his grandson Chapman S. Root learned he had inherited the family business while serving in the military during WWII.

Descendants of Chapman J. Root donated their private collection of Coca-Cola memorabilia to the Vigo County Historical Society in Terre Haute. In 2017 the   Society was scheduled to move into a larger facility to better display its many collections, including the featured birthplace of the Coca-Cola

bottle exhibits. After 94 years of preserving the past, the old museum had run out of space, and many items had been stored out of sight.

To help make the new museum a reality the Root Family Foundation, in 2017 pledged a gift of $100,000. John Root, foundation board member, has worked closely with the Vigo County Historical Society. Museum representative , Susan Tingley said the center would continue to be "good stewards" of the Root family legacy in Terre Haute.

The museum is the historic Ehrmann building, located in the center of Terre Haute's downtown. The east wall of the building was planned to become the site for a giant mural of a Coca-Cola bottle, which may be the largest bottle in the world.

Photos and Research: Vigo County Public Library

The museum's entrance features a 1950's style soda fountain, which visitors can enjoy without paying the museum admission fee. Inside there are photos, posters, old advertisements and a vast collection of Coca-Cola items. There is a reproduction of the original Coca-Cola bottle mold, and a working bottling line constructed for the museum by Coca-Cola historians.

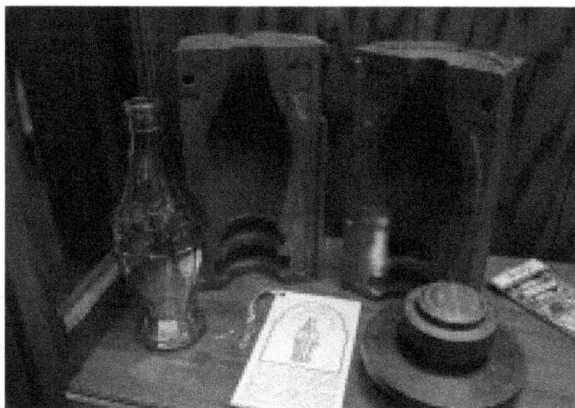

The city has joined the bottle's birthplace promotion with pieces of themed public art and murals throughout the downtown. New signage on the way into town proudly tells everyone Terre Haute has its own, exclusive Coca-Cola heritage.

When Chapman S. Root's home in Terre Haute was destroyed by fire, instead of rebuilding, he decided to move the company to Daytona Beach, Florida; where today the Root family museum provides an amazing record of more of the Root Coca-Cola experiences.

The Root museum opened in 2001 as a 24,000 square foot wing at the Museum of Arts and Science History in Daytona Beach. It is described as one of the largest Coca-Cola memorabilia collections in the world "with every conceivable item relating to the bottling, advertising and consumption of Coca-Cola". The

museum information explains, "the Root family has amassed one of the most historically important anthologies of the American soft drink on which the family fortune was founded."

The Root Organization also has, at its Ormand Beach office, an unique reminder of the relocation to Florida. In a special office display case sits a "droopy" Coke bottle, which was partially melted and retrieved after fire destroyed the Root Indiana home.

There are two 1925 Coca-Cola delivery trucks on display. Another unique exhibit allows insight into the lives and travel of early American industrialists. Two mid-century train cars, which were converted by Root into luxurious private rail coaches, are displayed in an enclosed train station.

Three race cars, represent the eight years when Root entered his Sumar race team in the famed Indianapolis 500. The cars attracted considerable attention be-cause of many new "streamline" design ideas. The best finish was a sixth place in 1953, the team's rookie year in competition. Two Sumar drivers were killed in the final years, and Root closed the race shop after saying racing was no longer fun. The orig-inal prototype of the 1915 Coca-Cola bottle

Photos Credit: Museum of Arts & Sciences, Daytona Beach

was placed on display in the museum's Root Family Wing in November 2015, to recognized the bottle's 100th anniversary. It is the only prototype bottle on display in the world. Another prototype is in a vault at the Coca-Cola headquarters in Atlanta. Also donated to the museum were the original patent documents for the bottle.

Another prototype bottle was sold at auction in 2011 for $240,000. The bottle had been in the possession of Brad Dean, grandson of Earl Dean who was one of the original bottle designers at Root Glassworks. The auction was held in Beverly Hills, California, but the final destination of the rare bottle is not known.

As their Coca-Cola business evolved, Root and his wife Susan continued to acquire Coca-Cola artifacts; building a collection which included bottling equipment, posters, trays, the famous Coca-Cola calender girls, and a wide variety of other items with Coca-Cola logos.

By 1982 Root's empire, Associated Coca-Cola Bottling had plants in St. Louis, Pennsylvania, Florida, New York, New Jersey and the Virgin Islands. Root's health was failing, and he also became unhappy with some Coca-Cola corporate decisions, including the introduction of "new Coke".

Coca-Cola was buying many independent bottlers at that time, and the decision was made to sell Associated Coca-Cola for $417.5 million. The Root family owned 57-percent of the company and netted $238 million from the sale.

Chapman's son, John learned of the sale when he called his parents from Hawaii to tell them he was getting married.

John later said, "I was in total disbelief at the time. It had always been part of our family. There is such history there that I never really thought we'd get out of it."

When Chapman S. Root died in 1990 he left behind a family-run company which had diversified into communications, real estate, exports and philanthropy. Preston Root is President of the Root Organization with a primary interest in real estate. However, the non-profit Root Foundation proudly continues the family tradition of community support and participation, with a special emphasis on the critical needs of children and families.

An article in the Orlando newspaper described the Root family as "a silent angel" in the community. "They operate their charitable trust like an invisible hand that makes an imprint without leaving fingerprints." The foundation has provided over forty years of confidential donations totaling millions of dollars. Brothers Preston and John Root represent the foundation. They are great grandsons of Root Glassworks founder Chapman J. Root.

Three other siblings live in Idaho, which had been a favorite vacation spot for the family. Chapman J. Root has retired there. Christopher is in publishing, and William Root is a rancher. Their sister, Susan Graham lives in Hawaii where she runs a supply company.

Just like the Fords and the Rockefellers, the Roots have been described as a "classic American family".

Chapter 6
Corinth, MS

A Coca-Cola partnership in the early 1900's between two businessmen in north Mississippi provided the catalyst for the future generations of two families committed to the bottling and distribution of Coca-Cola. The partners were A. V. Weaver of Corinth, Mississippi and C.C. Clark, originally from Martin, Tennessee.

Weaver was the grandfather of Kenneth and H.L. "Sandy" Williams, brothers who own and operate Corinth Coca-Cola Bottling Works, where four generations of family members have worked during the company's more than 100 years in business.

Clark was the grandfather of Harold Clark, who is Vice President of C.C. Clark Incorporated of Starkville, Mississippi, a beverage business where three third generation and six fourth generation family member are actively involved.

In 1903 C.C. Clark was operating a soda water business in the rear of his drug store in Martin when he decided to establish a small soda water bottling plant in Corinth. Meanwhile Weaver, who was selling pants for his brother's pants manufacturing company in Corinth, had seen a bottling plant during a sales trip to Arkansas.

Weaver became interested in the bottling business and in 1905 purchased half interest in the Corinth plant from Clark. At that time the small plant had three employees. News about Coca-Cola caught Clark's attention and the two partners began efforts to obtain a bottling franchise from Ben Thomas in Chattanooga. Coca-Cola was being distributed in Corinth at that time by a wholesale grocery company. It was being shipped by train from the Coca-Cola plant in Jackson, Tennessee. However, Weaver and Clark soon were able to take over the area distribution for the Jackson bottler.

In 1906 while waiting approval to bottle Coca-Cola in Corinth, Weaver and Clark were granted a license for a bottling plant in New Albany, a community about forty miles south of Corinth.

Clark moved to New Albany to begin that operation. He hired the post-master, W. G. McGill as his manager, and McGill ultimately purchased the business in 1911. The McGill family continued to operate the New Albany plant until 1986, when once again it became a part of C.C. Clark, Incorporated.

Weaver and Clark expanded to West Point, Mississippi in 1906 when each acquired one-third interest in a bottling plant for that community.

The Coca-Cola franchise for Corinth was granted by Thomas in 1907. Soon after Weaver and Clark reached a new agreement which resulted in Weaver controling Corinth and New Albany, and Clark becoming owner and operator at West Point.

It was believed that Paul Russell from the Jackson Coca-Cola plant had been influential in making the franchise available for Corinth, and he became a significant stockholder in the Corinth operation. Four others also were involved in minority ownership in the early years of Corinth Coca-Cola. However, by 1940 Weaver became sole owner when he bought out Paul Russell, the last remaining shareholder.

Corinth's first bottling plant was a simple operation with bottles washed and filled by hand, and capped by a foot powered machine. Distribution to rural areas was made by rail, while businesses in downtown Corinth received their Coca-Cola orders from a strong young man with a wheelbarrow. Eventually the very early gas powered trucks were placed into delivery service.

An on-going inventory of glass bottles was a critical issue for the early Coca-cola bottlers, as was apparent in an invitation sent out in 1919 by Corinth Coca-Cola. The company announced plans for a July 4th open house event, and residents were invited to visit the plant.

Corinth Coca-Cola bottling plant 1916 on Waldron Street

Bottling room at Waldron Street plant. Man on right is Albert Weaver, brother to company founder A.K. Weaver

Inside the plant 1910, small boy holding bottle is A.K. Weater Jr., about 6 years old at the time.

However, the invitation also emphasized the need of the Corinth plant to have an adequate supply of empty bottles, not only for the open house, but also to fulfill the larger orders anticipated for the July 4th holiday. The notice asked "friends to help out" by making a special effort to ship back cases of empty bottles every day.

The local bottlers inventory of glass bottles was one of their largest investments, and reuse of the bottles was financially important.

A new plant was constructed at the current location of 601 Washington Street in 1949.

Corinth Coca-Cola observed its' 100th anniversary in 2007, and collectors will find it interesting to learn that a limited edition of 2500 Coca-Cola trays were produced as part of the celebration.

"Sandy" Williams remembers how some basic marketing research was done in the earlier years. It was a time when sodas were sold at retail outlets from a vending unit which had a bottle cap remover attached. The Coca-Cola route men would collect the used bottle caps from the

Coca-Cola moves to new location on Washington Street in 1949

machines, and when they returned to the plant the caps were sorted and counted to determine the percent of sales for each brand.

Sandy and Kenneth in 1982 with portrait of their grandfather, company founder Avon Kenneth Weaver.

Visitors to Corinth have the opportunity to view memorabilia from the Coca-Cola plant. An exhibit was first created by the Corinth company as part of the 100th birthday celebration. A variety of vending machines, advertising, signs and other items were displayed along with a replica of an old fashion soda fountain. All items remained on display in a special museum building, until a flood in 2010 damaged the building forcing it to be closed.

About a year later a decision was made to place many of the old Coca-Cola items back on display at Corinth's popular Crossroads Museum, which has been described as "a show case for Corinth's rich past".

The museum is located in the city's historic train depot at 221 N. Filmore Street in downtown Corinth. The museum is called "Crossroads" because it's the place where the Memphis & Charleston and the Mobile & Ohio railroads crossed, creating a strategic transportation hub during the Civil War. The museum is open Tuesday through Saturday, 10am to 4pm.

Another Coca-Cola tradition in Corinth is the annual 10K race, held every year in May. Kenneth Williams founded the race in 1982 to encourage and reward physical fitness in the community. Many of Corinth

Ken and Sandy Williams with one of the Coca-Cola displays at Crossroads Museum.

Coca-Cola's 300 employees help organize and run the event, which will attract the race's maximum of 1500 runners.

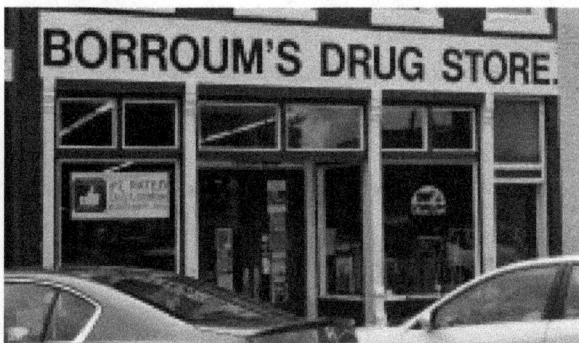

If you are thirsty after the race there is one more "must see" attraction in Corinth. You can enjoy a real soda fountain Coca-Cola at the oldest drug store in continuous operation in Mississippi. Borroum's Drug Store and Soda Fountain was founded in 1865 by a former Confederate Army surgeon,

A. I. Borroum. Dr. Borroum started the store as the war was ending, after being released from a northern prison camp.

The store is listed as one of the state's historic places. It features an old style soda fountain which serves Coca-Cola, milk shakes, and other drinks. And if you are hungry the store is known throughout the area for its' signature "Slugburger". Served daily, the tasty burger's recipe is said to include ground pork, soy meal, flour and salt.

Corinth Coca-Cola is one of five U.S. Bottlers which were selected by the Coca-Cola Company to receive new, expanded distribution territories. In 2015 Corinth expanded its presence in west Tennessee by obtaining territories in the Jackson and Paris, Tennessee markets. The acquired area reaches from the Tennessee River west to the Mississippi River, and north to the Kentucky state line.

Corinth in north Mississippi is a worthwhile stop on "The Coca-Cola Trail".

GULF STATES CANNERS

Sandy Williams of Corinth Coca-Cola is one of the state's local bottlers who also were instrumental in founding another Mississippi Coca-Cola enterprise.

Similar to the glass bottle needs of that July 4th in 1919, another packaging need was becoming a concern for bottlers in the late 1960's. It was the growing need to provide Coca-Cola in cans, as consumers were rejecting returnable bottles in favor of the convenience of throw away cans.

Some thought serving this trend for canned Coca-Cola might be a matter of survival for the local bottlers. However, the equipment and staff needed to establish a canning line was too expensive.

While moving forward with a solution to this new packaging challenge, Williams also reflected on the earlier days of the iconic glass Coca-Cola bottle. He remembers how each bottle was identified with the plant's name on the bottom. Often when a group of friends were enjoying a Coke, there would be a contest to see whose bottle was from the plant furthest away.

The bottlers also calculated the average number of times a bottle would be reused. When that number did not work out, and more bottles were needed, it might be thought that another bottler was using someone else's bottles.

But now it was time to meet the new challenge. Customers were no longer looking for plant names on a bottle, they were seeking throw-away convenience.

To solve the problem, Williams and eight other Mississippi bottlers supported the concept of creating a beverage canning cooperative. Research was done, a corporation was formed, and within two years a plant was producing Coca-Cola in cans for its' Mississippi owners.

It was one of the nation's first independent beverage canning co-ops. Gulf States Canners in Clinton, Mississippi is said to be a "national model of container excellence".

Chapter 7
Aliceville, AL

Much has been written about Coca-Cola's success in establishing bottling facilities to provide Coca-Cola to American soldiers in Europe during World War II. However little has been told about a Coca-Cola plant in the United States which often operated 24 hours

Photo courtesy of Aliceville Museum

a day so that much of its' production would be available for German prisoners of war and the U.S. Soldiers who supervised them at a large POW camp in Alabama.

The history of both the Coca-Cola plant and the POW camp are featured in a one-of-a-kind museum in Aliceville, Alabama. The city owned museum is probably the only one in the country to actually be housed in a former Coca-Cola plant. Visitors to the plant have the opportunity to see some of the plants' original bottling equipment, dating back to 1948 and earlier. Other museum displays feature Coca-Cola memorabilia, historical records and photos.

The Aliceville plant began operation in June of 1910. It was owned by David Bowling as part of a franchise agreement from Sam Kaye of Columbus, Mississippi. Kaye Coca-Cola Bottling Company continued for several years as an active distributor in Mississippi.

Beginning in 1914 a series of ownership changes took place during the following eight years.

Ted Martin is listed as owner in 1914, but his involvement is brief, as the next year the plant is co-owned by Sam Summins and J. T. Hardin. The plant changed owners again in 1917 when it was acquired by three business men; James Murphy Summerville, Hugh Summerville, and E. J. Pierce. In 1922 George Downer purchased Pierce's interest in the plant. Downer eventually became sole owner, and continued to operate and manage Coca-Cola bottling in Aliceville until 1954.

Downer oversaw the plant's operation during the demanding production requirement to provide Coca-Cola for the POW camp. Sugar supplies to bottler was rationed during the war. But it was said Aliceville was granted special allotment consideration because of the commitment to the camp. A photo of Downer at his desk was included in an article about the

Photo courtesy of Aliceville Museum

Aliceville plant, which was featured in the November 1949 issue of "The Coca-Cola Bottler" magazine.

It also was during Downer's ownership that a new plant building was constructed at the same location. The cost for the new plant and equipment in 1948 was reported to be five thousand dollars. That building and adjacent warehouse and vehicle garage ultimately became the current home for the city museum.

One of those who helped build the plant in 1948 was a young student, James "Jimmy" Summerville, now residing in Chico, California. Jimmy was in high school when hired in 1946 to work on the "soaker" on the bottling line. He also recalls helping deliver Coca-Cola to the prison camp. Jimmy continued his part time and summer plant employment while studying at Auburn University.

Jimmy became a brick layers' helper in the summer of 1948 as the new plant was being built. He estimated he must have handled at least half of the number of bricks used for its' construction.

James Summerville

He graduated from Auburn in 1953, but values his Coca-Cola memories, and visits the old bottling line whenever he returns to Aliceville.

The plant changed hands again in 1954 when it was sold to Paul Watts. During his ownership a new annex structure was completed, and the plant's 50th anniversary was observed in 1960 with a special event attended by over 700 people.

The Aliceville plant experienced two more ownership changes. Three partners; J.V. Park, Jr., Carl Chandler, and Paul Martin were listed as owners in 1973. Finally consolidation occurred when the Aliceville plant and its distribution territory were acquired by the Meridian, Mississippi Coca-Cola Bottling Company. The Aliceville operation was shut down, but Meridian continued to use the building for storage until 1993.

At that time the Aliceville Museum was housed in the public library, and in need of more space. Hardy P. Graham, Chairman of the Board of Meridian Coca-Cola notified the city the plant was being donated for the new museum. As part of the donation, it was agreed that the original bottling equipment would remain on display.

Museum Executive Director John Gillum is proud of what may be the only museum of its' kind in the country. He explains his Coca-Cola exhibit contains the complete and intact bottling equipment, just as it was installed when the plant was constructed. Included is a large Miller-Hydro bottle washer and a 20 valve filler and crowner. Some of the machinery on display had been relocated from the original plant and was in use before and during the war.

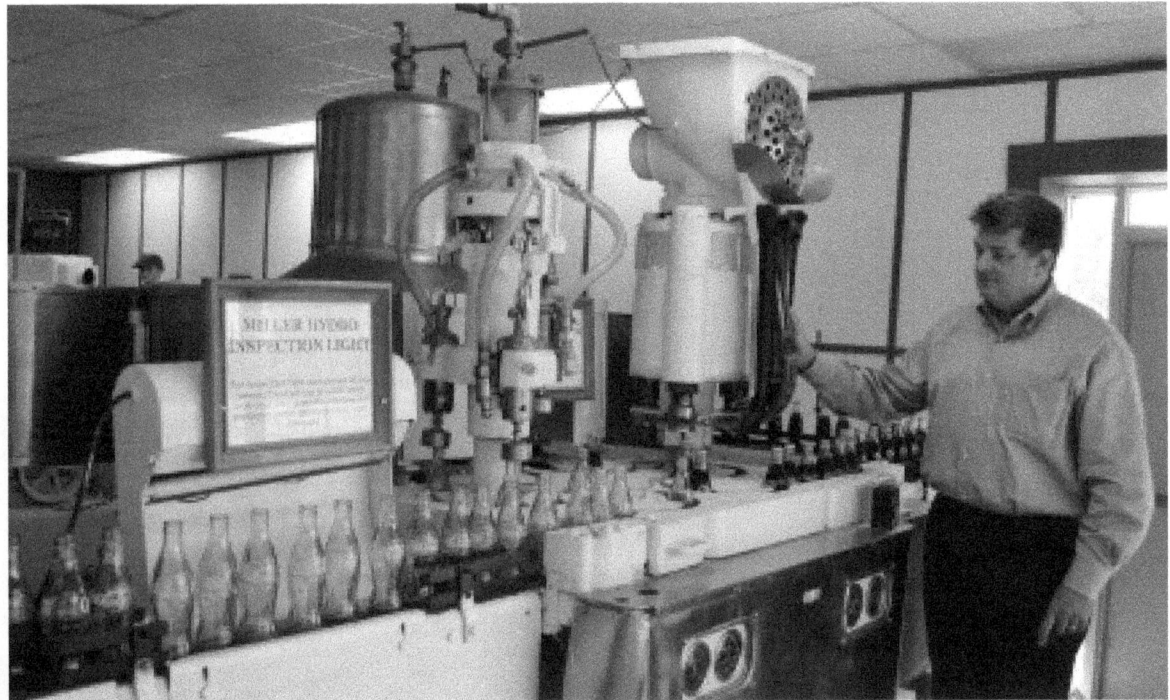

Gillum points out that visitors to the museum also are able to view the largest collection in the United States of World War II prisoners of war items. Included are works of art and various items created by the Aliceville prisoners, along with clothing, documents, and other camp artifacts.

Another museum display is described as "the American Heroes Collection", with military items dating back to world war I.

The Aliceville Museum is located at 104 Broad Street. The phone number is 205-373-2363.

Chapter 8
Unique Coca-Cola Signs

Outdoor Coca-Cola signs are much more than a form of American product advertising. Through the years the aging signs often survive to become beloved murals reflecting a special time or memory. Many times the signs become a treasured part of a community's history.

Sometimes individuals or groups around the nation will see the historic value of the old signs ,and will take action to save, restore and preserve them.

This "stop" on the Coca-Cola Trail reveals some of the various ways the signage has been saved, and we pay tribute to the on-going efforts of the Coca-Cola mural preservationists.

Coca-Cola signs always will be "the real thing".

Cartersville, GA

Outdoor signs were among the first forms of advertising for Coca-Cola. They were relatively inexpensive, which was important for a struggling new company with limited promotional funds.

In 1886 an oilcloth banner was affixed to an awning at Jacobs Drugstore in Atlanta. The banner's message was simple, "Drink Coca-Cola 5¢" It became the first point of sale advertising for Coca-Cola.

The world's first Coca-Cola outdoor wall sign was painted in 1894 on the side of Young Brothers Pharmacy in downtown Cartersville, Georgia. It was created by Coca-Cola syrup salesman James Couden, who had been making a sales call, and asked store owners W.W. and G.W. Young for permission to paint the sign on the building's east wall. The sign has been authenticated by the Coca-Cola Company as the first outdoor wall sign.

The pharmacy building was constructed in 1811. W.W. Young became involved in the business in 1890, and his brother G.W. Young joined him a few years later.

A project to have the sign restored was initiated by Dean Cox, who had purchased the business in 1968, and then bought the building two years later. Cox first asked the Coca-Cola company to do the sign restoration, but his request was turned down.

Not giving up on his project, Cox located two lady sign painters in the area who agreed to take on the unique task. They used electric heat guns and hand scrapers to carefully reveal the outline and remains of the original sign. Four months of detailed work by the two painters was required to completely restore the sign in 1989.

The sign received recognition from the Georgia Trail Awards in 1990. A media event featuring the restored sign followed, and attracted international attention, including TV coverage by all three major broadcast networks, as well as cable networks including CNN.

Cox related the history of the sign during media interviews, and later a Coca-Cola representative remarked that the brand must have received over one-half million dollars worth of publicity.

Cox later revealed that Coca-Cola not only decided to pay for the sign restoration, but provided him a bonus as well. The sign is now listed on the National Register of Historic Places.

THE FIRST PAINTED WALL SIGN TO ADVERTISE COCA-COLA WAS PLACED ON THIS WALL FOR THE COCA-COLA COMPANY IN 1894. THIS SIGN WAS RESTORED IN 1989

If you visit the pharmacy, look closely at the letter "i" in the word "drink". Cox explained

it appears to be slightly smaller and squeezed into the word. The reason he believes, is the original painter first forgot to paint the letter "i", and then had to add it to the word "drink". That 120 year old mistake remains in the current restored sign.

The Young Brother Pharmacy sold soda fountain Cokes until 1945. Today its inventory includes a variety of Coca-Cola memorabilia items which can be purchased by the thousands of fans who visit each year. The store maintains a journal where visitors from throughout the world have signed their names.

Coca-Cola realized early the importance of outdoor advertising, and it was estimated that by 1914 the company had created over five million square feet of painted walls. Coca-Cola President Asa Candler once boasted in Hollywood "it would be difficult to make a movie outside without catching a Coca-Cola sign in the background".

The first Coca-Cola outdoor neon sign was erected in 1929 in New York's famous Times Square. The sign measured 75 x 100 feet and at the time was the second electric sign in the world. It flashed the message "Drink Coca-Cola, Delicious and Refreshing". That sign was moved to 47th street in 1932 and featured a soda jerk in Coca-Cola uniform.

Coca-Cola has maintained a Times Square location for more than eight decades.

In July 2014 the Company unveiled a new sign in Times Square to commemorate the 100th anniversary of both Times Square and New York's Coca-Cola bottler. Once again the sign is one of the largest in the world. It features a three-dimensional high tech display and is more than six stories high. It required the work of more than 40 engineers and designers to create the new sign. It weighs 30 tons and has over 2.6 million LED lights, operated by a state-of-the-art computer system.

Kearney, NB

A restored sign in Kearney, Nebraska brings back memories of the long forgotten Coca-Cola "Sprite Boy" cartoon figure. The Sprite boy was introduced in the 1940's, but has not been used to promote Coca-Cola for over fifty years. But now, here he is, restored for all to enjoy once again on the side of the Coca-Cola building at 119 W. Railroad Street in Kearney.

Coca-Cola created the elf-like cartoon figure in response to a marketing problem. People had begun using the term "Coke" for the soft drink, a practice the Coca-Cola Company at first tried to discourage. However, the company finally gave in to the will of their customers and Sprite Boy was unveiled in June 1941 to acknowledge and promote the "Coke" nickname. Coca-Cola legally took ownership of the nickname when it registered "Coke" March 27, 1944.

Sprite boy was created by artist Hadden Sundblom, who also created Coca-Cola's still popular Santa Claus images. Sprite boy and Santa appeared together in an advertisement in 1943 which supported U.S. troops and encouraged Americans to buy war bonds. The two were together again in a 1949 Christmas advertisement entitled "Travel Refreshed". In the ad Sprite boy watched the reindeer while Santa paused to enjoy a drink of Coca-Cola.

Sprite boy would be drawn wearing one of two different hats. When the advertisement featured bottled Coca-Cola his hat would be a bottle cap. When the message featured Coca-Cola in a glass, Sprite boy wore the cap of a soda fountain jerk.

The Sprite boy campaign was created by Archie Lee of D'Arcy Advertising company. Sprite is derived from the Latin word "Spritus". The Sprite beverage was introduced by Coca-Cola in 1961, many years after Sprite boy had been retired.

The rare wall sign in Kearney was restored in 2008. It is located on the side of a Coca-Cola facility owned by The Chesterman Company of Sioux City, Iowa. The Kearney plant was purchased by Chesterman in 1985 from Grand Isle Coca-Cola in Hastings.

Chesterman, a pioneer in the U.S. Bottling industry, was founded in 1872 in Dyersville, Iowa by Cilo Chesterman, to bottle soda waters. His son, B. Chesterman started bottling Coca-Cola in 1904.

In addition to its bottling operation in Sioux City, the company has Coca-Cola facilities in Aurelia, Iowa; Sioux Falls, Yankton, Mitchell, Pierre, and Watertown, South Dakota; and MaComb, Illinois.

Opelika, AL

Restored Coca-Cola wall signs always attract attention and are valuable as they preserve the history of the marketing of the most recognized brand name in the world.

However, an original non-restored sign of more than 100 years in age is a true treasure. This piece of rare Coca-Cola history was uncovered in June 2014 at a hardware and building supply store in Opelika, Alabama.

Store owner Dozier Smith T had started a remodeling project when he discovered the 100 year old secret. His family has operated the business for three generations, but no one knew the sign was there.

Dozier had removed an old wooden panel when he saw the letter "A" along with "5 cents". Old plaster covered the remainder of the sign and he became concerned that removing the plaster would damage the sign. Dozier proceeded with patience and caution and the plaster started to come down. It took two months before the complete floor to ceiling sign was revealed.

It turned out the original sign was in excellent condition. Originally it had been painted on an outside wall adjacent to the hardware store location. However, construction of the hardware store building in 1910 resulted in the sign being covered after only about two years of outside exposure.

The wording on the sign"Relieves Fatigue" was important to Coca-Cola historians in determining the age of the sign. That particular phrase was used in Coca-Cola promotion between 1907 and 1912. The Opelika sign is believed to be the oldest original, non-restored Coca-Cola sign in existence.

Dozier reflected that possibly his grandfather, Winston Smith T, may have been the first in the family to see the sign. Winston, who founded the family hardware business in 1931, would have been about six years old when the sign was painted, and Dozier believes he probably saw it.

The sign has become a tourist attraction in Opelika attracting thousands of Coca-Cola enthusiasts.

An official unveiling ceremony for the sign was held October 9, 2014. Several hundred people turned out and each received a nickle which was used to purchase a bottle of Coca-Cola. Opelika Mayor Gary Fuller repeated a familiar Coca-Cola slogan, "I'd like to buy the world a Coke". Coca-Cola representative Andy Britton praised Dozier and his family for their efforts to preserve an important piece of Coca-Cola history.

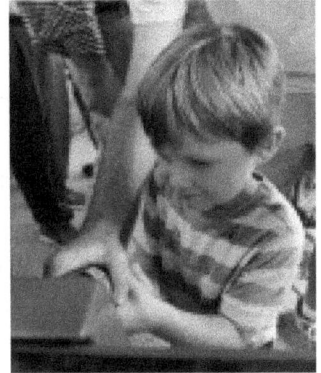

Coca-Cola was first bottled in Opelika in 1902 by George Curtis Roberts, who had started his soda bottling business about ten years earlier. The franchise was obtained from Joseph Whitehead in Chattanooga, with the first bottling plant located at 1st and 8th street. Robert's grandon, Curtis Roberts Vance, believes the bottling actually may have been done in the current Smith T Hardware building, or in an adjacent structure; which would explain the location of the recently uncovered original sign.

A new Coca-Cola plant was built in Opelika in 1938 at 7th and Railroad Streets. A newspaper article described it the "latest design in a handsome building". An open house was held at the new facility May 26, 1938, with residents invited for "refreshments and souvenirs". It is interesting to note a newspaper notice invited "colored people" to visit the plant the next evening, May 27th from 7:00 pm to 9:30 pm.

The plant was sold in 1975 to become part of William A. Bellingrath's Montgomery based operations.

The plant was sold again in 1990 to Coca-Cola Corporate, and bottling continued in Opelika until 2007. Then, after almost 70 years of operation, the decision was made to move to a plant in West Point, Georgia, about 18 miles away, and all 43 employees at Opelika were able to transfer to the new location.

Soon after it was closed, the abandoned plant caught the attention of Lisa Ditchkoff, a Bostonian who grew up in Oklahoma, but "fell in love" with Opelika and had moved there in 2001.

She and business partner Dale Downing saw a new future for the old Coca-Cola plant, and after a lengthy battle to obtain government SBA funding, they were able to purchase the building in 2008.

The partners created a downtown event center for Opelika, and the Coca-Cola building came to live again as a place for weddings, business meetings and other community activities.

Ironically the building was the site for a meeting of Coca-Cola United employees from Birmingham the day before the celebration for the old sign at the Smith T Hardware store.

Lisa sees the building as a way to help in the development of downtown Opelika.

In 2016 the building went "back to its roots", as it partnered with Coca-Cola United to become the "Bottling Plant Event Center".

Executive Chef Jonathan Walker said, "We are trying to keep the history of the building alive."

Baton Rouge, LA

Coca-Cola bottlers often will do whatever is required to save and restore old outdoor signs in their community. However one of the most challenging, but important, of these restorations is a large lit roof top sign in Baton Rouge, Louisiana. The sign is said to be one of only three of its' type in the United States, and the only one still in use.

In 2015 a dispute over the actual ownership of the sign threatened its' continued existence. The sign first was erected by the local Coca-Cola bottling company in 1951 on the roof of Liggett Rexall drug store in downtown Baton Rouge.

The drug store closed in 1981, but the sign remained and was refurbished by the bottler in the mid 1980's. Other businesses occupied the location through the years, and the sign was refurbished again in 2002, and finally once again in 2014 at a cost of $20,000.

In recent years the building was owned by Elvin and Joycelynn Richoux, who operated a family restaurant there. When the Richouxs decided to sell in 2002, they donated the sign to the Baton Rouge Arts Council with an agreement that insurance and electrical costs for the sign would be paid by the downtown business association. The building was purchased by Danny McGlynn, and he acknowledged in 2013 that he did not own the sign, when the building was again sold to Mike Crouch of MDC Properties.

In May of 2014 a chicken restaurant chain, "Raising Canes", opened an outlet on the ground floor of the building. One day before the restaurant's opening, building owner Mike Crouch had the sign covered with a black tarp. He was demanding an agreement from the business association and the arts council which included paying for insurance and maintenance of the sign, as well as a "market rate" advertisement fee.

After several days, which saw the issue gain media and public attention, Crouch agreed to remove the tarp covering, but the sign remained dark for over a year. The sign is unique because it is illuminated by large neon tubing, and believed to be the only one of three such signs still in operation.

The dispute finally was resolved in August 2015, and the sign was re-lit. The Baton Rouge Area Foundation assumed responsibility for the sign's maintenance and operation.

An attorney representing Crouch said his client was satisfied with the final agreement, which made possible the preservation and enjoyment of Baton Rouge's historic Coca-Cola sign.

1906. The first Baton Rouge Coca-Cola Bottling Co. plant, corner of St. Louis and Europe Streets.

Coca-Cola was first bottled in Baton Rouge in 1906 when businessman Thomas Daigre opened his plant in a small downtown building. The city's population of 12,000 people received their Coca-Cola via Daigre's one horse wagon distribution system.

Today Baton Rouge Coca-Cola is a division of Coca-Cola Bottling United of Birmingham, the nation's third largest independent bottler. A modern new bottling plant began operation in Baton Rouge in 2008. It is one of three bottling facilities operated by United to serve their 35 distribution centers in seven southeastern states.

Monroe, LA

The growing interest in old outdoor Coca-Cola signs provided the spark for a tourist promotion campaign in Monroe, Louisiana. Not just one, but four fading outdoor Coca-Cola murals caught the attention of the local convention and visitors bureau, in an on-going project to promote the downtown areas in the twin cities of Monroe and West Monroe.

Monroe also is home to the Biedenharn Coca-Cola Museum, a tribute to Joe Biedenharn, the first person to bottle Coca-Cola, and the Biedenharn family which became one of the nation's largest Coca-Cola bottling company.

Visitor's bureau executive director Alana Cooper saw the aged outdoor signs' potential for a cross marketing effort to attract visitors for both the museum and the downtown areas.

Two of the murals are located in the popular Antique Alley, along the west side of Ouachita River, which divides Monroe and West Monroe.

Local artists Brooke Foy and Emery Thibodeaux took on the restoration project. The four murals first were pressure washed. Then the artists contacted the Coca-Cola archives to obtain exact color images for each sign.

Thibodeaux said she had admired the signs since she was a child, and gave an enthusiastic "yes" when presented with the challenge of restoring them.

All four murals can be considered rare, as each one features the Coke "Sprite Boy", an elf-like cartoon figure not used in Coca-Cola advertising since 1950.

The community provided support for the project, and all four were completed in five weeks.

One of the murals is located on the former Monroe Coca-Cola bottling plant on the east side of the river.

Another is located along Louisville Avenue, a busy Monroe street which leads to the river bridge.

Sponsors of the restoration project included Coca-Cola Bottling Company United, the current Monroe plant owner; the Biedenharn Museum, and area civic and economic development organizations.

A ceremony to recognize the refurbished murals was held, with Louisiana Senator Mike Walworth cutting the ceremonial ribbon. The senator also is a former Coca-Cola employee, and he commented that the restoration was another excellent way to create attention and help uplift the community. Others present for the ceremony included Monroe plant manager Keith Biedenharn, and museum president Hank Biedenharn.

Coca-Cola Twins

Two old original Coca-Cola murals, one in Tennessee and one in Virginia are good examples of how "historic treasures" can remain undiscovered for generations. In addition, those two signs could be classified as "identical twins". Both signs, although hundreds of miles apart, were painted by the same sign company in 1906.

Both had been placed on outside building walls until new construction covered them, to remain gone and forgotten for over 100 years. And most amazing about those Coca-Cola twins, both were discovered when each location was being renovated to house a restaurant.

In Morristown, Tennessee the sign had been painted on the Myers Dry Goods store. But only a year later the store was expanded and the sign was covered. The three story building was housing a TV repair business in 1999 when it was purchased by Kirk Fenner, and work got under way for a new restaurant and

the "Higher Grounds" coffee shop on the upper level. Little by little, the sign, which was covered by stucco, was carefully revealed to become a feature of the

restaurant. The sign's size required a 40' x 18' cutout on the third floor to allow for a full view of the Coca-Cola treasure.

The Morristown restaurant is located on the corner of Main and Cumberland, on what has been called the "white lightening highway", as it served a popular moonshiner's route from Knoxville to Johnson City. Future plans for the building include the addition of a moonshiner's museum to chronicle those colorful days in the city's history.

Additional Coca-Cola memorabilia in Morristown can be enjoyed at the nearby "Meeting Place and Country Store". Owners John and Terry Springer have created an ice cream soda fountain on the first floor, and John's Coca-Cola collection is proudly displayed in the second floor Country Store.

The "twin" to the Morristown mural is located 480 miles away in Suffolk, Virginia. Both signs had been painted by "The Gunning Systems" of Chicago, a sign company at that time well known for creating outdoor advertising on the sides of large structures.

The Suffolk sign was discovered by Harper Bradshaw when he was renovating an old building to create Harper's Table Restaurant. The building had been vacant for about a decade when Bradshaw bought it in 2011 and began a major project to bring it up to code as a commercial structure. The building last had been used as a pawn shop, and in the 1970's was a paint store. The original site for the building was an alley, and in 1906 the sign was painted on an adjacent structure. It is believed Harper's building was constructed soon after, resulting in the sign being covered and hidden from view.

Harper had been removing wood paneling from a wall when he discovered the amazingly preserved mural. He quickly realized the sign would become a centerpiece for his new restaurant, and design plans were changed to accommodate the historic attraction. The sign was cleaned

carefully with a special vacuum. No restoration was needed, and a clear coat was applied to preserve and protect.

The owner of the building's 1970's paint store became a regular customer at Harper's table, and often remarked he never knew his store contained a "hidden treasure".

The historic Coca-Cola sign "twins", remain proudly featured at two southern restaurants, thanks to the efforts of two alert businessmen.

Lafayette, Colorado

A large Coca-Cola mural, which almost was lost forever, has a new life thanks to the dedicated efforts and financial support of the residents in Lafayette, Colorado.

The 14' x 26' sign was discovered in the summer of 2015 when workers were demolishing an old business building which had been vacant for over six years. The sign was hidden behind some exterior siding.

The sign had been painted in the 1930's on the side of the Hi-Way Restaurant, along U. S. highway 287, now known as Public Road. At that time highway 287 was routed through the city. The sign was repainted in the 1950's when the business became Pat and Gar's Hi-Way Bar, and may have been covered by the siding when the business became a Mexican restaurant, La Familia.

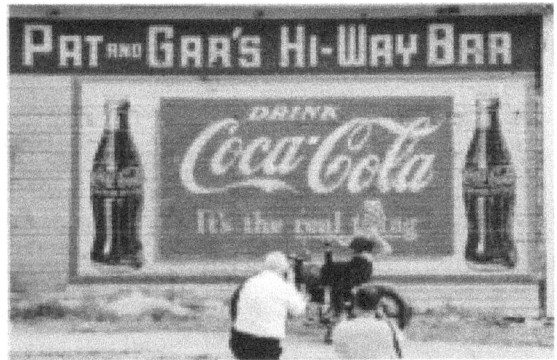

Crews demolishing the building had been asked to be alert for anything historic. Lafayette Urban Renewal Authority Chairwoman, Sally Martin explained it is their goal to preserve that which is Lafayette history.

When the sign was discovered, a special mural committee was established to determine the future for the Coca-Cola sign. The city took the first step when $10,000 was authorized to detach and save the 4300 pound sign. Lafayette had established a popular public art streetscape program.

Photo, City of Lafayette

The mural was stored at the fire station for eighteen months, while the committee launched a mural fund raising project. While in storage the sign received attention from local conservationists Lisa Capano and Deborah & Uhl, who spent several months repairing and conserving the mural. Meanwhile Lafayette citizens and businesses showed a strong endorsement of the project by donating over $162,000.

Next the city located a permanent home for the sign, which would be on Public Road only one block south of its original location. On November 30, 2016 the sign was transported from the fire station and installed on an historic building owned by Richard Ross, a local developer. Ross had purchased the building a year earlier to restore it and create a business with his nephew.

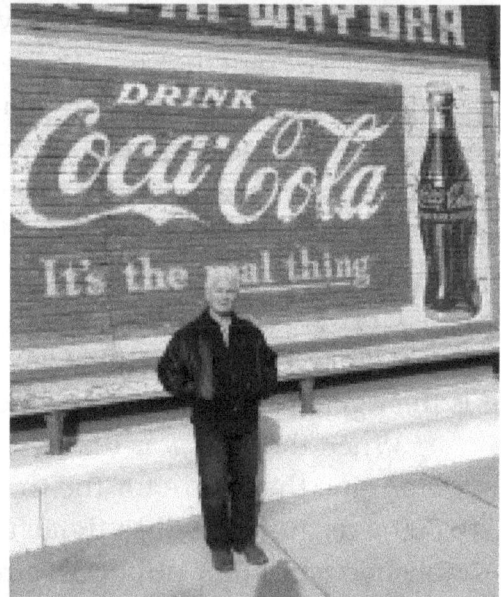

His new business, The Black Diamond Restaurant and Tap House, opened a few months later. The business name honors the Black Diamond mine, which was located in the Lafayette area when the city was a major coal town, for over fifty years beginning in the 1880's. Two other businesses share the restored building, a home décor store and a shop operated by an eighth generation chocolatier.

The building, which had been constructed in 1900, throughout the years had been home to a variety of businesses, including a bank, a pawn shop, cafes and the Sportsman's Bar. Some original paintings, done by an artist in exchange for drinks at the bar, were discovered in the building and now are on display at the Black Diamond. Also on display is the original vault from the First National Bank.

After the installation, the sign was dedicated December 3rd during a city sponsored ribbon cutting event, to recognize the importance of preserving another part of Lafayette's history.

Among those participating in the event were building o wners Richard Ross and Renewal Authority Chair Woman Sally Martin

Chapter 9
Paducah, KY

A Coca-Cola plant in Paducah, Kentucky, once described by its owner as "not the biggest in the world, but the best" is living up to that claim attracting visitors and customers for the new businesses in the historic structure.

The plant was built in 1939 by Luther Carson, who had started bottling Coca-Cola in Paducah in 1903. The new plant was capped with a distinctive thirty-foot round dome made of copper, select maple, and translucent glass brick. Colored lights were placed behind the glass and operated by a clock mechanism with color affects at night. The foyer featured a marble floor with a large Coca-Cola logo in the center. There was a 10-foot chandelier in the

shape of a Coca-Cola bottle, and a spiral stair-case leading to second floor offices.

In an interview in 2003, which marked the 100th year of Coca-Cola in Paducah, Carson's daughter Jane Myre remembered her father's dedication to the building's construction.

An old bottling machine on display in the plant lobby was demonstrated by Luther Carson

Photos Credit: From the collection of the McCracken County Public Library Special Collections.

She said, "he watched almost every brick that went into it. He once sent back a railroad car of lumber because the wood had too many holes."

It also was said Luther hand-picked every piece of wood to be used in completing the second floor offices. No expense was spared when he built his plant.

Carson first became interested in bottling Coca-Cola when he and his brother John were working on their uncle's incline mountain train in Chattanooga. There they met bottling rights owners Thomas and Whitehead, who used the train to reach their homes on Lookout Mountain. The brothers decided to invest $2700 for a Coca-Cola franchise covering a 65-mile radius of their home town of Paducah.

They started their new bottling business in a rented two story building on south Third Street. The first case was produced at 10:35 am, March 27, 1903. Luther took the case to nearby Wolf's Grocery and told owner George Wolf he wanted him to have the distinction of buying the first case of Coca-Cola bottled in Paducah.

Wolf was an old friend, and not to disappoint Luther, bought the case for 70-cents. He placed the case on the sidewalk in front of the store to see if customers would try the new Coca-Cola. At first people were just curious, but eventually all the bottles were sold and Wolf became convinced and placed a standing order with Paducah's Coca-Cola Bottling Company.

To get started the Carson brothers would bottle at night and use mule drawn wagons to peddle Coca-Cola during the day. The brothers paid themselves a monthly salary of $60 each, and the plant netted $250 during the first nine months.

A year later the Carsons moved to a larger plant at 5th and Jackson Street, and also took their first step into expansion, when they purchased the Coca-Cola franchise for Evansville, Indiana. John left Paducah to become manager at Evansville, about 180 miles north.

Sales continued to increase in Paducah and three years later the plant was moved to a larger building, just a block away, at 6th and Jackson. Bottling continued at that location for thirty years until the 1937 Ohio river flood which filled the city streets and surrounding areas with high water and devas-

Photos Credit: From the collection of the McCracken County Public Library Special Collections.

tation. The flooding stretched from Pittsburgh to Cairo, Illinois leaving hundreds of people dead and over one million homeless.

Carson escaped his flooded plant by using a syrup keg to float from a second floor window. After being pulled from the water by a rescue boat, Luther commented, "If I ever reach dry ground it is there I will build my new plant."

Plans for his new state of the art bottling operation began immediately after the flood. The plant is located on "dry ground" on land across the street from the former flood evacuation site, where Luther had been taken after his rescue. A micro brewery called "Dry Ground Brewing" began operating in the old plant in February 2015.

Luther Carson was 89 years old when he died in 1962, after sixty years of being a Coca-Cola bottler. The family continued to operate their interest in eighteen Coca-Cola plants in four states. But by 1984 they believed "the day of the small bottler over". Several potential buyers considered the business, including Marvin Herb who had large plants in Indianapolis and Chicago. The Carson business was sold in 1986 to Coca-Cola in Chattanooga; a bit of irony as Chattanooga was the place where the Carson brothers first became interested in bottling Coca-Cola.

Paducah developers Ed and Meagan Musselman bought the building in 2013, after it had sat empty for seven years. In bringing the building back to life, the developers worked to preserve the historic value, while creating "something new and exciting for the community". Dry Ground Brewery became the first tenant, and others include Mellow Mushroom pizza and Pipers Tea and Coffee. The work of local artists is showcased as well.

A lifesized mural depicting the Coca-Cola plant's opening night celebration in 1939, has been painted on Paducah's river wall by renowned mural

artist Robert Dafford. The artist also has painted Coca-Cola murals in Vicksburg and New Orleans. The Paducah mural is one of more than fifty Dafford has painted on the Ohio River wall, called "Portraits from Paducah's Past". The public art experience has been named one of Kentucky's most popular tourist attractions.

Paducah also has honored their Coca-Cola bottler and his family with the "Luther F. Carson Center". Built in 1995, the performing arts center was named in honor of Carson in 2004 to recognize the Carson family's support of the center and arts in the community.

Chapter 10
Placerville, CA

The history of bottling and Coca-Cola in Placerville, California is unique as it includes two stone structures, both listed on the national records of historic places. Placerville is an old California gold rush days town, located about 45 miles east of Sacramento. Originally a tent city in 1848, it was first called "Old Dry Diggins". In 1854 it became Placerville, named after the Placer gold deposits found in the nearby rivers and hills. It is the county seat for El Dorado County.

The structure called the John Pearson Soda Works building at 594 Main Street, was built in several stages, with the first construction in 1859 by John McFarland Pearson, a Scottish immigrant. Pearson previously had operated as an ice merchant at his Silver Creek Ice Depot on Bedford Avenue. To construct this new location he purchased what was called the "old Fountain House" site on Main Street. Apparently he selected that site because it included an old mine which went back into the hill for 150 feet, and would maintain a cool year round temperature, perfect for ice storage.

Later Pearson expanded into the soda water business, and in 1884 a traveling salesman from Chicago was quoted as praising Pearson's Creme Soda as "the best in the United States".

The building took on some added uses when in 1867 it was used as a drill room by the Placerville city guard, and in 1874 it served as the armory hall.

John Pearson died in 1891, and his sons John and William inherited the building and continued their father's business. They added a second story to the building which housed the growing bottling operation. Also added was a water powered elevator to provide product transportation to and from the upper level. Soon after, the store began to sell basic grocery items such as butter, syrup and cider; as well as wines, liquors, imported ales and domestic beers including Pabst.

The soda works equipment and stock was sold in 1904 to Julius Scherrer and his brother, who rented the building and changed the name to Placerville Soda works, and remained in their rented facility until 1911.

The business in the building changed hands several times, but the Pearson family retained ownership of the building until 1972, when it was bought and restored by Roger Douvres. A lunch house and soda parlor were opened on the lower floor in 1978, and a dinner facility was opened on the upper level in 1978.

The Cozmic Cafe occupied the building in 2015, and utilized the old mining tunnels as unique seating areas. The cafe offers foods, micro brews, a variety of coffees, wine and live music. It boasts of being "Placerville's hub for the arts with a strong emphasis on community".

When the Sherrer Brothers moved their bottling business in 1911, they had constructed a wooden structure on Main Street, located across from the present office of the El Dorado Chamber of Commerce.

Photo: El Dorado County Historical Museum

Included in the Soda Works purchase by the Sherrer Brothers was the franchise for Coca-Cola, but apparently they did not think it worthwhile to produce Coca-Cola.

Photos: El Dorado County Historical Museum

When Robert A. Hook bought out the Sherrer Brothers in 1934, he began to produce Coca-Cola, much to the Sherrer's surprise. Hook and his son continued to run the bottling company until it was sold in 1967. The plant had moved into a new facility in 1960, and was using the old wooden structure for storage until it was destroyed by fire a year later.

The Placerville Coca-Cola franchise was purchased by Coca-Cola Bottling of Sacramento, which at that time was owned by Nathan and Gladys Sellers. The Sellers had acquired the Sacramento franchise in 1928 for $200,000. When they purchased Placerville the bottling was consolidated to the Sacramento plant and Placerville was used for storage and distribution. Today the Sacramento franchise is owned by Coca-Cola Atlanta (CCR).

A second historic register building in Placerville also had served as a soda water plant. The rock structure built in 1852 became the home for the John Fountain and Benjamin Tallman Soda Works. Spring water was carbonated, bottled and sold to the miners, because the water from the nearby rivers and creeks had become polluted and unsafe to drink due to the mining activity.

The walls of the building are two feet thick, and there are iron shutters for the door and windows to provide protection from fires. Today the building is the Fountain & Tallman Museum, which is operated by the El Dorado County Historical Society. It has survived the fires and is the oldest building on Main Street, providing an example of the early permanent buildings which transformed Placerville from a mining camp into an actual town.

In addition to its unique history in soda bottling, Placerville also provides visitors a look at other memories from the gold rush days.

One of those memories is of John Studebaker, a name familiar to many because of the Studebaker automobile legend. However, Studebaker first built a "vehicle" of a different type after arriving in Placerville in 1853.

He had journeyed from his home in Indiana to seek his fortune in the gold fields. However a friend suggested he might have more success by providing items needed by the miners. With that thought in mind he rented a small work shop in the rear of the Ollis and Hinds Blacksmith shop. For five years he repaired wagon wheels and manufactured wheelbarrows for the miners. That effort earned him the nickname of "Wheelbarrow John", along with an accumulated savings of some $8,000.

Photo: El Dorado County Historical Museum

John took his savings and returned to the family business in South Bend, Indiana where he joined his brothers in making farm wagons. Later the name became even more famous when the Studebaker auto was produced.

An historical marker was placed in 1939 at the site in Placerville where Studebaker had first worked. The ceremony was attended by hundreds of of citizens and distinguished visitors, including representatives of the Studebaker Automobile Company. The marker is located on Main Street, across from the office of the Chamber of Commerce. It has been recorded as site #142 on California's list of historic landmarks.

JOHN MOHLER STUDEBAKER
PIONEER BLACKSMITH-SOLDIER
INV... BUILDER
DEDICATED ... JAMES MARSHALL
CHAPTER NO. ... ...MPUS VITUS

Chapter 11
Mobile, AL

Bellingrath - The Gardens Coca-Cola Built

Bellingrath Gardens, a nationally recognized visitor attraction is described as "the charm spot of the deep south". The 65 acre site of year round floral pagentry is located along the Fowl River south of Mobile, Alabama. Included is the Bellingrath mansion with fifteen rooms and over 10,000 square feet containing original antique furniture and impressive collections of china, silver and crystal.

The gardens are the result of pioneer Coca-Cola bottler Walter Bellingrath of Mobile following the advise of his physician who had urged more relaxation and less business activity for his patient. In 1918 Bellingrath purchased the first 25 acres at the site, on which he planned to build a fishing camp.

Walter was the first of several members of the Bellingrath family who would become successfully involved in the business of Coca-Cola.

It began in 1902 when Crawford Johnson of Birmingham acquired the Coca-Cola bottling rights for most of Alabama for a fee of $3800. Johnson then sold territory which included Montgomery, Selma and Andulusa to C. V. Rainy and W. C. Cole. At that time Walter Bellingrath was operating a small brokerage company in Montgomery, while his brother William was in Anniston where he managed a commissary for the Woodward Coal and Iron Company. It was there William began to notice the growing popularity for the new Coca-Cola drink, as more of his customers were requesting it by name.

William and Walter decided together they would purchase the Montgomery Coca-Cola business from Rainy and Cole; and the plant, equipment and territory were acquired for about $5,000. Walter then closed his brokerage business so he could devote full time to their new plant, while William was finishing his employment contract with the Woodward Company.

Soon Walter made another purchase for the brothers, when Crawford Johnson sold them the Mobile franchise for $10,000. When William was able to become active in the business the brothers decided to split the operational duties with William remaining in Montgomery and Walter moving to Mobile.

Walter was 34 years old when he assumed responsibility for the Mobile plant, which also sold ginger ale, soda water, seltzer and mineral water. The total production of only a few cases per day was being done with a foot powered bottling machine.

The business had a mule drawn wagon and one helper. Bottling would be done one day, and the next day deliveries would be made as well as sales pitches to saloon and restaurant owners and potential retail customers. During those trips empty bottles were retrieved to be washed and filled the next day.

Walter had started the business with an unexpected bottle shortage, as he discovered he had received a smaller bottle inventory than he had been promised. Walter also used large metal cases so he could ship Coca-Cola and other products by rail and river boat to distant customers. Those sales provided a new and important source of revenue for the struggling bottler.

Continued hard work and long hours resulted in growth for the Mobile plant, and by 1905 daily production averaged 30 cases shipped and another 62 cases delivered in the city. The price was 75 cents for a 24 bottle case.

Nature delivered an unexpected blow to the Mobile plant in September 1906. Walter, an "inlander", had never before experienced a storm so violent; the worst to strike the city since 1893. He and a helper obtained cotton sacks and sand, and quickly made sandbags to position along ground floor openings in the plant. The storm pushed water to a depth of nearly three feet, and the building sustained wall and roof damage. Coca-Cola bottles and shipping cases were lost.

The first Mobile Coca-Cola plant immediately after the 1906 hurricane.

Walter & Bessie in door of plant a few years after the 1906 storm.

The plant was closed temporarily for several days. But in spite of the chaos, plans continued for a November wedding for Walter and Bessie May Morse. Bessie was a stenographer for Walter, and the two had become attracted to each other through the daily business activity. The couple honeymooned at Niagara Falls, and later moved into a home only a short distance from a new and larger Coca-Cola plant which Walter had purchased after the hurricane.

By 1910 Walter and William had been able to help create the Bellingrath family "Coca-Cola dynasty", which extended from Mobile into Arkansas. Their sister Kate and her husband W.N. Brown were operating the franchise in Selma. Another sister, Mary Elizabeth (Mamie) and her husband J. S. Burnett had the plant in Andalusia, Theodore Leon and wife Maude were operating the Little Rock, Arkansas plant, and Leonard Ferdinand and wife Jane ran the plant in Pine Bluff, Arkansas. All of this Coca-Cola enterprise had been created in just the first seven years Walter was in Mobile.

The second Mobile Coca-Cola plant in 1911

Walter's experience and enthusiasm for Coca-Cola bottling caught the attention of Charles Veasey Rainwater, who was Secretary-Treasurer of the Atlanta parent company. Rainwater assumed his leadership position after the 1906 death of Joseph Whitehead, who established the Atlanta company.

Working with Bellingrath and other bottlers was one of Rainwater's on going challenges, and he soon realized he needed the advise of "older and wiser heads". Walter Bellingrath was one of seven pioneer bottlers named to Coca-Cola's first advisory board.

The board first shared with others information on changes and technical progress in bottling. Then in 1909 the board held the first bottlers' meeting in Atlanta to discuss common problems and exchange ideas.

Walter became involved in another Mobile business in 1913, which ultimately would provide a needed product for all the Bellingrath Coca-Cola bottlers. The new company was the Lerio Patent Cup Company, which was built around the product idea of Louis Lerio, a master metal craftsman, who had moved to Mobile three years earlier. His product was a metal cup system used to extract raw rosen from pine trees, to be used in the manufacturing of turpentine.

Walter had a special interest in the product because it reminded him of his childhood days in Castleberry, Alabama where his father also had been involved in making products for turpentine manufacturers.

The Lerio metal fabrication facility provided a needed new product for the Bellingrath bottlers when it created a new style delivery truck body. The unit was successfully put into action in Mobile as well as on trucks at other family plants.

The Coca-Cola delivery truck created by Bellingrath's Lerio Company.

It was during his involvement with Lerio that Walter experienced his second encounter with a fierce gulf storm. It was July 1916, and both the Coca-Cola and Lerio plants were damaged along with flooding up to five feet in depth. Walter's home also received extensive roof and ceiling damage, and he was beginning to believe that for him Mobile was a "city of catastrophes".

Again looking to expand his business interest beyond Coca-Cola, Walter purchased the Mobile Ornamental Tile Company in 1920. The company's staff of artisans hand-made Mediterranean style tiles which were used in decorating private and public buildings. Some of their creations remain today in Mobile, New Orleans and other cities, to provide another reminder of the man named Bellingrath.

The Plant Built in 1911, with additions served as Mobile's Coca-Cola Plant until 1975.

During the 1920's Walter continued to attend Coca-Cola bottlers' meetings in Atlanta. He was recognized as a successful innovator for the production and marketing of Coca-Cola.

A writer, doing a feature on Walter's life in a magazine in 1951, made this observation, "He got into the Coca-Cola bottling business in 1903, which as it turned out, was somewhat like digging a shaft to a gold mine."

However it was those physical demands of business and community involvement by Walter, which became the catalyst for the creation of what would become Bellingrath Gardens. It was Dr. P. D. McGehee of Mobile who told Walter to get out into the country and get away from business. It is interesting to note the "good doctor's" grandson, Tom McGehee in 2015 was serving as Museum Director and Curator at Bellingrath Gardens.

Walter's original plans for his get-away site was to build a comfortable bungalow to use as a fishing camp and place to entertain a few friends. However, the camp quickly expanded to 60 acres when he bought adjacent property which included three old shacks in need of considerable cleaning and rehabilitation. Work quickly got underway and the first visitors were welcomed to the camp January 1, 1919.

Meanwhile Bessie was increasing her efforts to improve the camp's appearance. She obtained large Azalae and Camellia bushes from old homes in south Alabama, Mississippi, and Louisiana. She also visited well known gardens in South Carolina, Delaware, Pennsylvania and elsewhere. The Bellingraths traveled to Europe in 1927 where they toured famous English and French gardens.

Walter and Bessie first invited the public to share the beauty of "Bell Camp Gardens" on April 7, 1932. The response was much greater than expected, and foretold the future for a popular Alabama attraction.

It continued in February 1933 when Walter placed a newspaper advertisement inviting visitors to tour the "Bellingrath Gardens", as they now would be called, for a fee of fifty cents.

Construction of the 15 room mansion at the gardens began in 1935, and was completed in July 1936 when the Bellingraths enjoyed their first meal in the new house. Six months later they moved from their home in Mobile to become full time garden residents.

Bessie began to experience health problems in 1942, and Walter took her to Hot Springs, Arkansas for rest and treatment in 1943. However she died there unexpectedly February 15, 1943 at the age of 64.

Walter returned to his gardens, and on August 6, 1949 observed his 80th birthday by inviting all to visit and tour at no charge. Some 18,000 people accepted the invitation, and were treated to free Coca-Cola at refreshment stands along the garden paths. Walter Bellingrath died August 8, 1955 at the age of 86.

For more than sixty years the gardens and home that "Coca-Cola built" have survived hurricanes, financial challenges and the gulf oil spill. Thousands of visitors each year enjoy the year round gardens and special events, which include a spectacular Christmas lights display. The gardens are listed on both the Alabama Register of Landmarks, and the National Register of Historic Places.

Located in the Asian-American garden area a large round red sign, which is affixed to an Oriental style structure, reads in Japanese "Drink Coca-Cola".

The Bellingrath Foundation continued to own the Mobile Coca-Cola plant until 1981, when it was purchased by Wometco of Miami. At that time Wometco was one of the largest franchise bottlers with 14 bottling plants and an additional 25 distribution facilities.

Three years later the Wometco Coca-Cola business was purchased by Coca-Cola Bottling Company Consolidated of Charlotte, North Carolina, which is the largest independent Coca-Cola bottler in the United States. Bottling was continued at the modern Coca-Cola plant located at 5300 Coca-Cola Road in Mobile.

The Bellingrath name remained prominent in the Coca-Cola business in 2015 at the CCBCC corporate office in Charlotte. William Bellingrath Elmore was listed as Vice President. His uncle was Stanhope Elmore, whose sister Mary, had been the wife of William Bellingrath in Montgomery. Stanhope had assisted Mary in the operation of the plant in Montgomery after William died in 1938. Stanhope died in 1961.

Walter Bellingrath

Chapter 12
Fort Smith, AR

A Coca-Cola business in Fort Smith, Arkansas, which has been operated by four generations of the same family since 1895, also is the home to one of the nation's largest and most accurately chronicled private Coca-Cola museums.

It all started in 1892 when two brothers, J.W. and Robert Meek formed a partnership to operate a retail grocery in Fort Smith. Three years later the two brothers purchased the Fort Smith Bottling and Candy Company from J. D. Elliott. Their new business was located on the ground floor of a three story building on north sixth street.

In 1899 J. W. and Robert relocated the business to a new three story building on Rogers Avenue, and the company remained at that location for 89 years, although experiencing some building changes along the way.

One of the more unique structural challenges came in 1940 when the third floor was removed. The decision to down size to a two story building was made after watching a fire at a nearby three story structure. The Meeks observed the fire department's water pumpers were unable to reach the top floor of the burning building. Consequently as a precaution against a future fire, the decision was made to remove the top floor of the Coca-Cola plant. At the same time an addition was constructed to one side of the building to provide more office space and a garage. Like most Coca-Cola plants, the bottling operation was plainly visible thru large ground floor windows.

The Meek brother had obtained the right to bottle Coca-Cola in 1903 when a sub-bottler agreement was signed with M. W. Fleming of Little Rock, who had purchased bottling right for the entire state of Arkansas. The Fort Smith agreement granted a territory with a radius of fifty miles. Records show 190 gallons of Coca-Cola syrup were used the first year in Fort Smith.

The Meek brothers ultimately purchased the complete rights to bottle Coca-Cola from Fleming in 1907 for a fee of $1500. Their new territory included a larger section of western Arkansas, as well as some counties in eastern Oklahoma.

Early Coca-Cola bottling in Fort Smith

In 1911 the Meeks established sub-bottler agreements in Paris and Rogers, Arkansas and in Poteau, Oklahoma. Another sub-bottler was established two years later in Harrison, Arkansas. Many of the sub-bottlers were short lived ventures, and Meeks ultimately obtained and operated the plant in Poteau.

The Coca-Cola territory in Muskegee, Oklahoma was purchased by the Meeks for $30,000 in 1917, and a sub-bottler was contracted in Okmulgee, Oklahoma.

During this time period unauthorized bottling of Coca-Cola was becoming a problem for many of the franchised bottlers. One example was discovered in Clarksville, Arkansas where a bottler was purchasing fountain syrup from a local wholesaler for use in his own bottling plant. Robert Meek, who had become aware of the Clarksville operation, notified Coca-Cola in Atlanta. A subsequent threatening letter from Atlanta, put an end to the illegal bottling.

During the following growth years for Fort Smith Coca-Cola more sub-bottler agreements were signed, including Fayetteville, Bentonville, and Eureka Spring, Arkansas; as well as Tahlequah, Stilwell, and Stigler, Oklahoma.

Additional beverage lines also were bottled and distributed thru the years, including Orange Crush, Green River, Delaware Punch, Sprite, Fanta, A & W and Barg's root beers, and Schwepps.

In the 1920's area retailers could purchase Coca-Cola at the Fort Smith plant for resale in their business. The cost of Coca-Cola, when picked up at the plant, was 30-cents for a case of 24 bottles, plus a deposit of 50-cents.

The plant's first delivery trucks were purchased in 1915. Previously deliveries to towns outside of Fort Smith had been done by rail, with shipments in barrels of 18 dozen bottles, or in cases of three or six dozen bottles.

A 1927 Mack delivery truck has been located and refurbished by the company. Now with a fleet numer of

"Coca-Cola 1", it has been seen in parades and at community events and promotions, including an important Wal Mart Corporation visitor center dedication in Bentonville.

After 89 years the move to a new plant in Fort Smith was completed in 1981, to the current location on Phoenix Avenue. The site had been acquired a few years earlier, and some of the company's divisions already were operating there. The local historical society earlier had expressed an interest in the new plant site, as it also had been the location of the original Fort Smith.

Throughout the history of Fort Smith Coca-Cola the business always has been managed by direct descendants of James Stanhope Meek, who was the father of the company founders James and Robert Meek. Now the fourth generation of managing partners are cousins Roger Meek, whose great grandfather was James; and Bob Hunter, whose great grandfather was Robert.

Managing partners Roger Meek and Bob Hunter with restored delivery truck.

The idea for a company Coca-Cola museum came from employee Fred Kirkpatrick, who in 2015 had been with the plant for over 77 years, and would boast of having worked with all four generations of managing partners. Fred was semi-retired in 2015, but remained involved as manager and curator of the museum.

Fred had been collecting Coca-Cola memorabilia on his own, but became interested in creating the museum after attending the second annual convention of the Coca-Cola collectors in 1976. That event had been held at the once famous, but now closed, Schmidt Museum in Elizabethtown, Kentucky.

The Meek Museum of Coca-Cola memorabilia was established in 1988 with four display cabinets of well documented items. The museum grew to feature 30 display cases, as well as larger items such as vending machines and old bottling equipment.

Fred Kirkpatrick with museum displays.

Two unique museum displays are the result of extensive research and acquisitions. Proudly displayed on museum walls is a complete set of Coca-Cola issued calendars beginning with the year 1914. Also on display is a 1903 calendar, which has special meaning, as that was the year the company first started bottling Coca-Cola.

A second unique display is a collection of sports related Coca-Cola commemorative bottles. Fred's efforts to locate the bottles resulted in a display of 600 of the sports memories.

Many of the museum's items came from the regular course of doing business thru the years. However others are rare items of special interest, which were located and purchased.

In addition to his museum duties, Fred also maintains the company's archives room, which contains records and historic documents, including the original 1907 Coca-Cola bottling rights from M. W. Fleming.

Coca-Cola Bottling of Fort Smith observed its' 100th anniversary in 2003, and at that time commissioned well known Arkansas artist John Bell to create a painting of the plant as it appeared in 1916, when the transition was being made from horse drawn wagons to gasoline powered trucks.

Bell lived in Fort Smith and was famous for his "hometown series" of paintings which featured the city as it appeared around the turn of the twentieth century. The framed Coca-Cola plant original, along with prints of all others in the series, are proudly displayed in the company's conference room.

The artist, who died in 2013, created his work while confined to a wheelchair and painting while battling the mobility problems of Cerebral Palsy. His many honors include the 1971 President's Award for distinguished service to persons with Cerebral Palsy. The museum is located on the second floor of the Coca-Cola building, and often is used by organizations visiting Fort Smith. The museum is available by appointment.

The actual bottling of Coca-Cola was discontinued at the Fort Smith facility in 1990. However, the line was put back into operation one more time in 1992, to create a commemorative bottle for the city's sesquicentennial; another example of the Meek family tradition of support and participation in Fort Smith, Arkansas.

Chapter 13
Grenada, MS

Grenada Mayor Billy Collins and a close friend, together made it possible to preserve the memories of a Coca-Cola plant which operated in their central Mississippi city for over fifty years.

The Coca-Cola Bottling Company of Grenada was started by H.L. Honeycutt in the 1920's. Honeycutt also served on the city council and was a well known, respected resident of the area.

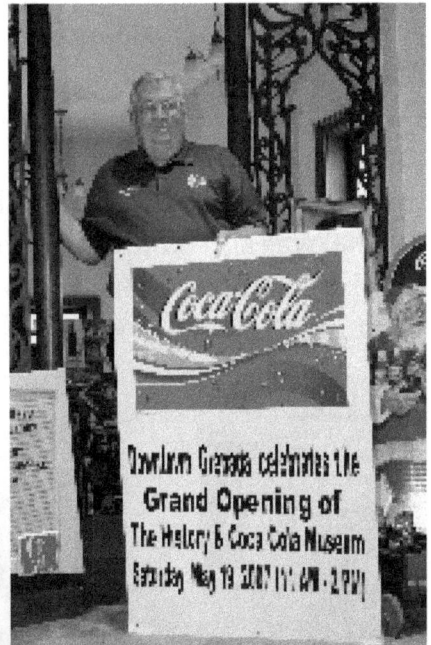

Coca-Cola Plant
1941

His son Louie H. Honeycutt continued the operation of the plant for several decades after his father died in the 1940's. And true to Coca-Cola tradition, when Louie died in the 1970's, his daughter Kathleen assumed control of the plant until it was sold in 1976 to Greenwood, Mississippi Coca-Cola.

Kathleen had married K. Veasey and brought with her a house full of Coca-Cola memorabilia from the old plant. Upon her death in 2005 the Coca-Cola collection was included in the estate left to her four children.

Soon it came to Mayor Collins attention that the children were planning to auction off the items on the internet. The news of the

Bottle washer removed from closed Coca-Cola Bottling Plant.

pending sale had been brought to the mayor by his old friend Ray Oliver, and both agreed the history of Grenada Coca-Cola should not be allowed to leave the city. Oliver, who was a member of a successful area business family, also had a distant Coca-Cola connection, as his son had married one of Kathleen Veasey's daughters.

Together the Mayor and his friend approached Kathleen's son Larry, about the possibility of acquiring all of the plant's memorabilia. A reasonable price of $15,000 was agreed upon for the total collection. Then began the Mayor's challenge to raise the money, as he was determined to make the purchase without seeking city funds.

The Mayor's initial request for donations quickly raised $5,000. Ray Oliver then paid the remaining $10,000, which resulted in Mayor Collins then initiating a second search for donations to re-pay Oliver. Coca-Cola in Atlanta provided assistance, and ultimately the remaining amount was obtained from private donations.

While fund raising, Mayor Collins also faced the task of finding a home for the Grenada Coca-Cola treasures. The solution, quickly reached, was city hall at 108 S. Main Street, where the Mayor's office and adjoining meeting and work spaces became the temporary location for the Grenada Coca-Cola Museum.

The museum was open to the public May 19, 2007 and has become a popular attraction for Grenada visitors.

The Mayor said, "We have hosted Coca-Cola fans from throughout the nation, and we are making plans for a new and larger museum."

The city owns the Masonic Temple building, also located on Main Street, just a block away from city hall. Funds were being raised to remodel and convert the 15,000 square foot building to become a permanent home for Grenada's unique Coca-Cola Museum.

Visitors to Grenada also can enjoy a search for their own Coca-Cola treasures at the Vendors Antique Mall. Located just across the street from the former Masonic Temple, the mall features several displays of Coca-Cola memorabilia, along with other collector's items.

In addition, the old plant which housed Coca-Cola of Grenada is also located on Main Street, just a block north of the city square. Private businesses now occupy the building.

A mayor, a friend, and a generous city all made it possible to create another interesting location on The Coca-Cola Trail.

Chapter 14
Rocky Mount, VA

A Coca-Cola plant built in 1929 in Rocky Mount, Virginia has become a focal point in the revitalization of Rocky Mount's historic downtown area.

The former Coke plant, now a popular restaurant, actually was the second location for Coca-Cola bottling in Rocky Mount. The first site was a small rented building, where in 1921 brothers Harry and Moton Menefee started their bottling business. That first plant was only 20 x 70 feet in size, which required the loading of delivery trucks to be done outside.

The new plant, a two story structure, provided inside space for vehicle loading. The actual bottling process was unique, as the Coca-Cola was mixed in the

upper level, and then gravity fed to the ground floor where bottles were filled. It has been said the process gave the Coca-Cola a distinctive flavor. Other sodas, including grape, orange and ginger ale, also were bottled there.

Photos Courtesy Franklin County Historical Society

Coca-Cola sales grew and by the 1930's a local restaurant called "The Hub" had become the largest local Coca-Cola customer purchasing as much as 100 cases per week. However, sugar rationing during world war II forced the plant to reduce production, and the Hub received only 25 cases per week, and most of the area stores were limited to 5 cases each.

Another interesting point in the plant's history came when the Menefee brothers decided to stop using the glass bottles which had the Rocky Mount location imprinted on the bottom. Some say the decision was made because they learned of a competition common at that time, where vending machine customers would wager to see who would receive a bottle from the plant furthest away. Others believe cost may have been a factor in the decision, as glass bottles without the bottom imprint were less expensive. In any case, the plant owners later decided to add the name of their town to the bottle cap.

Moton's sons, Charles and Richard worked in the plant as young men, and later worked for, and retired from, the Coca-Cola Company. Harry died in 1947, and Moton continued to operate the business until 1952, when it was purchased by Coca-Cola Roanoke, with the plant to be closed and bottling moved to Roanoke. A new distribution facility was constructed in Rocky Mount, and used thru the 1960's.

Coca-Cola Roanoke is part of Coca-Cola Consolidated of Charlotte, North Carolina, which is the largest independent Coca-Cola bottler in the United States. When the decision was made to restore the Rocky Mount plant building to become

Bootleggers Cafe, Jack Farlin of Best Bet Arts in Roanoke, along with an assistant, restored all the old Coca-Cola signs in four days.

Farlin has done several other sign projects for the Charlotte bottler, which has made a priority of locating and restoring old Coca-Cola signs throughout its' vast territory.

Bryan Hochstein bought the former Coca-Cola plant building in 2013 to create his Bootleggers Cafe. Previously a pizza place had occupied the building along with apartments on the second floor. That upper area now is a dining hall.

Located across the street from the cafe is the Harvester Performance Center, which provides a site for entertainment and other events in a restored historic building. Hochstein cited the success of the performance center and the rebirth of the downtown area as important factors in his decision to open the cafe.

He said, "You want to be where the people are, and I believe Rocky Mount will be that place. New memories will be made while taking the past along for the ride."

The cafe owner said his restaurant is bringing back home-style cooking with fresh vegetables and fresh prepared foods.

Also located near the cafe is an historic Greek revival style plantation home which now serves as a popular bed and breakfast facility. Called "Early Inn at the Grove", the home was built in 1854 on a ten acre site, and is attracting new visitors and friends for Rocky Mount.

A ribbon cutting ceremony for Bootleggers Cafe was hosted July 9, 2014 by Coca-Cola Bottling Company Consolidated. The cafe at 467 Franklin Street, officially opened for business a few weeks later.

The Franklin County Historical Society Museum in Rocky Mount, has many artifacts from the time the Coca-Cola plant was in operation. Included is the original scales used at the plant to weigh sugar used in the bottling. Also at the museum are several display cases filled with rare advertising and other memorabilia.

The museum's special projects coordinator Lindy Stanley explained more items are in storage, and each year the museum features a seasonal display of Coca-Cola Christmas items.

Museum Secretary Doris Eames with
old Coca-Cola Christmas ornaments

The museum is located at 460 Main Street.

Chapter 15
Minden, LA

"A one-man recreation commission" for the city of Minden. That was the description of Coca-Cola bottler Larry Hunter in 1938 in this north Louisiana community.

But the history and commitment of Coca-Cola to Minden goes back much further. It can be traced to 1886 when Larry Hunter's father, William S. Hunter of New Orleans arrived in Minden as a Western Union telegraph operator.

A few years later William became interested in the growing popularity of bottled soda water, and in 1901 he founded the North Louisiana Bottling Works. Four years later he began bottling Coca-Cola as a sub-bottler for the Biedenharn Shreveport plant.

When William died in 1919, his son Larry and Larry's wife Gladys, took over the operation of the plant and began a fifty year commitment to community involvement in Minden.

A new plant had been built on the original site at 412 Pine Street in 1914. And again at the same location, the third and current facility was constructed in 1926. Also in 1914 Hunter decided to build a second plant in Homer, about twenty miles northeast of Minden.

Five years later the area experienced the beginning of an oil boom, with the first producing well in the Homer oilfield. Additional wells quickly followed in Lisbon and the Haynesville district; with the oil being produced there boosting Louisiana to third place in the nation's oil producing states. The new oil industry spurred a population growth and Homer grew to become as large, or even larger than Minden.

To meet the growth a new Coca-Cola plant was built in Homer in 1920, and it was remodeled and increased in size in 1948. After the oil boom gave way, the Homer plant ceased operation in 1972 with production being consolidated in Minden.

Meanwhile in Minden, Larry Hunter had recognized the need for recreation facilities in the city, and he and his wife set out to meet that need with a dedication which ultimately would provide a park, a swimming pool, baseball stadium and continued support for the youth and sports activities.

The swimming pool was built in the early 1930's on the Hunter property. It was followed in 1935 by a playhouse with adjacent playground equipment. The baseball park constructed in 1938, reflected Larry's love of baseball and sports in general. Not only did he construct and light a baseball park, he also brought in a top amateur baseball team, the Minden Red Birds, which at that time played in the Big Eight League. Hunter also provided team travel and paid road trip expenses for players on the American Legion ball team.

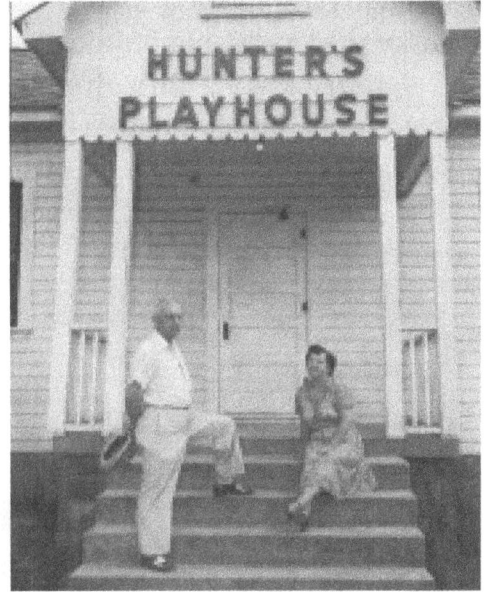

During the 1940's the Hunters realized many of the youngsters who earlier had enjoyed activities at the playhouse, now were serving in World War II. To welcome these young men back home, a larger playhouse was built complete with a dance floor and game rooms. And it was not uncommon in those years for Larry to load his big Buick with youngsters and head out for a trip to the state fair or a beach outing in Galveston.

By the end of 1948 the recreational facilities had become so large and popular they were becoming difficult for the Hunters to operate. After the urging of a local civic club, the city agreed to take over, and what followed was the first city supervised youth activities, and the beginning of a recreational program in Minden.

The Hunters were honored by Minden as citizens of the year in 1969. Larry died February 6, 1971 at the age of 74, and Gladys died two years later. A college scholarship was established in their memory and is awarded each year to a Minden high school student.

A radio broadcast by B. C. "Pop" McDonald in Minden in 1971 paid tribute to the Hunters. "Pop" said, "there is no way of measuring the value of Hunter's playhouse to the city of Minden. The swimming pool, the ball grounds and other things provided by the Hunters; things the city didn't provide at that time. Larry Hunter seemed to have the ability to bring out the best in young people."

An appropriate memorial to Larry and Gladys is located on the grounds in front of the Coca-Cola building.

Larry's son, Bill Hunter was involved in the family business for over fifty years, and served as President from 1950 to 1973. His brother, Ben Hunter lead the business from 1973 to 2001.

Don Hunter with the Hunter's memorial and an old bottling machine.

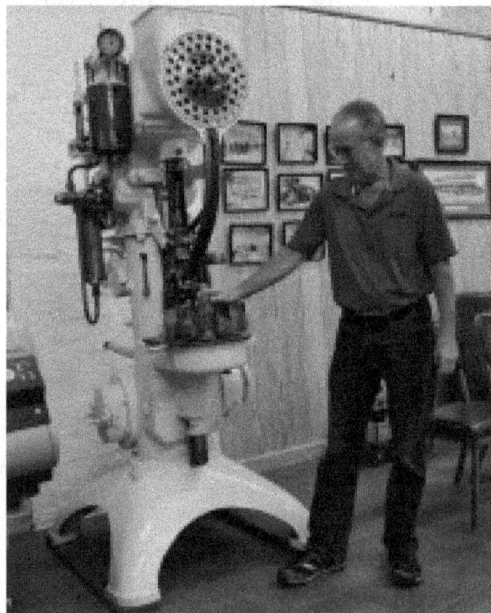

Ben Hunter in office with framed photo of his father Larry Hunter.

Outdoor sign in downtonw Minden originally pained in the 1980's.

A fourth generation of Hunters followed when Larry's grandson Don Hunter and Don's cousin, Jess Robertson assumed the management. Don retired in 2016, with Jess continuing to manage the family business. Before he retired, Don proudly reflected on the Coca-Cola commitment to the area, which was started by his grandparents and continues generations later.

Coca-Cola Minden continues to support numerous school activities, as well as area festivals and civic organizations. Product often is donated to school functions, with the schools keeping all funds raised by the sale of those donations.

Much of the Coca-Cola memorabilia from the Minden plant can be viewed in a special tribute exhibit at the city's Dorcheat Museum at 116 Pearl Street.

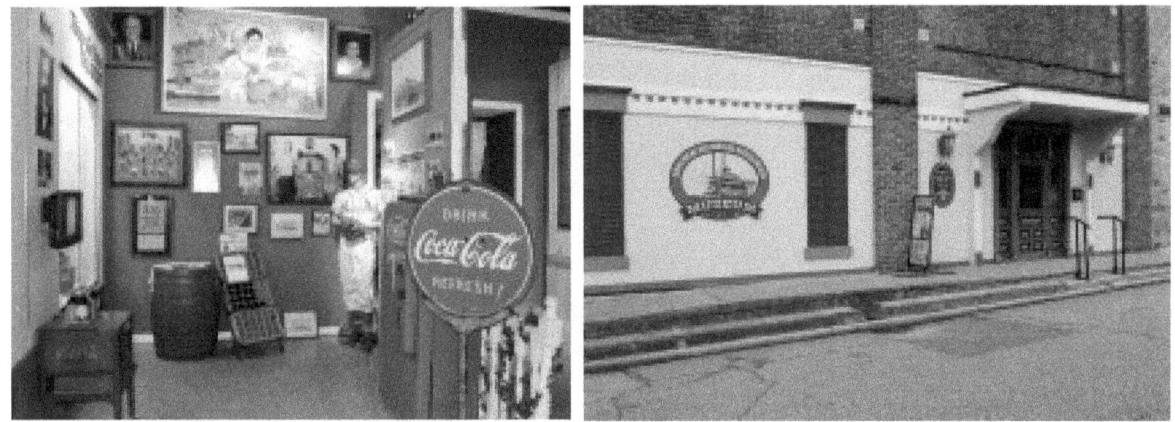

For the memorabilia shopper, a large variety of Coca-Cola items always are available at Second Hand Rose Antiques, 509 Main Street. Owner Milly Rose will provide a story for all of the items, along with an ice cold Coca-Cola, and she has been known for often singing her customers favorite songs.

Minden Coca-Cola provides a modern tribute to the earlier days of Coca-Cola bottling. It is Louisiana's only single unit operation still owned and managed by the original founding family.

Chapter 16
Cedartown, GA

A young man, who has been an avid Coca-Cola memorabilia collector since he was an eighth grade student, now has his collection on display in a former Coca-Cola bottling plant in Cedartown, Georgia.

Daniel Morris was nineteen years old when his Coca-Cola museum opened to the public June 25, 2016. Daniel's parents, Jammie and Darcy Morris had purchased the long abandoned plant in 2014, and the transformation to a Coca-Cola museum became a major family project.

The building was in such disrepair that the city once considered condemning it. A costly roof repair first was required. A conference room and museum offices were constructed on the second floor. On the

first floor, in the former bottling area, large display cases filled with memorabilia are featured. In the rear area where delivery trucks once were loaded, visitors now view displays of vending machines and other large items, as well as two old Coca-Cola trucks, a 1926 Model T and a 1954 pickup.

Daniel and his parents Jammie and Darcy Morris

Daniel proudly explained that he had located and purchased 98-percent of the memorabilia on display. His on-going search for "Coca-Cola treasures" has taken him to collector shows, antique dealers and flea markets in several states, as far west as Arizona and north into Indiana. One very rare item discovered at a show in Orlando, may be the world's first coin operated vending machine. It was made in 1912 by the Stanley Goddard Company in Atlanta, and the unit proudly displayed in the museum, may be the only one of its kind still in existence.

Another highlight of the museum is an 18-foot imported soda fountain which had been located and purchased in England. The classic, ornate fountain now greets visitors at the museum's entrance.

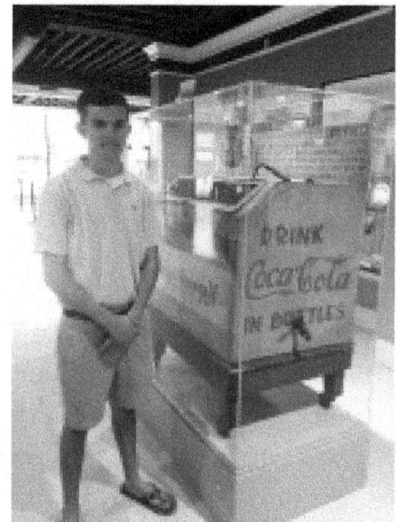

A theater room on the first floor presents a variety of old Coca-Cola films, and a 25-foot long wall of art features Coca-Cola history through the 1970's.

Visitor parking for the Cedartown museum is not a problem, as the Morris family purchased an adjacent lot. Jamie Morris explained the museum had received requests from group tours and school field trips, and realized the need for increased parking to accommodate buses. The museum also allows the lot to be used for overflow parking for the local performing arts center.

The Cedartown Coca-Cola plant was built in 1920 by F.S. Barron of Rome, Georgia. Barron created his own Coca-Cola empire with seven plants in Georgia. His first was built in 1901 in Rome, and at that early date probably was the sixth franchised plant in the nation. Other Barron plants were located in Valdosta, Fort Valley, Carrolton, Dalton, and Cartersville. The Cedartown plant was closed in the 1970's, and in 1986 the Barron family sold all its territory and remaining plants to Coca-Cola Enterprises of Atlanta for $84 million.

The Cedartown Museum of Coca-Cola is located at 209 Main Street. Public hours are 10am to 3pm Thursday and Friday, and 10am until 4pm on Saturday.

Chapter 17
Covington, TN

Coca-Cola bottling plants often became a focal point and source of community pride as more than 1200 of the independent family plants were in operation by 1930. As the popularity of Coca-Cola grew, so did the need for larger and more modern bottling facilities. Often the new plants provided an opportunity to showcase new trends in architecture.

One of the most unique plants was constructed in 1940 in Covington, Tennessee. It was designed by Memphis architect Everett Woods, and featured the Art Moderne style which was popular at that time.

Thanks to the efforts of a Covington firm, Rose Construction, Inc., the classic building was saved and has been placed on the National Register of Historic Places. Today the building serves as corporate headquarters for Rose Construction, a business founded in 1953 by the Rose family, as a small blacksmith shop. Rose Construction today is licensed in 14 states providing construction

services for commercial, industrial, and civil projects. The building has retained much of the original Coca-Cola style and features, along with a small museum featuring Coca-Cola memorabilia.

Beverage bottling actually began in Covington in the 1890's when Leander Baltzer operated the Covington Bottling Works. Baltzer acquired the rights to bottle Coca-Cola in 1909 from J. D. Pidgeon of Memphis, who earlier had purchased the larger territory franchise. Baltzer and Pidgeon formed a partnership and created the Coca-Cola Bottling Company of Covington. Pidgeon was president but continued to reside in Memphis, where he also established Memphis Coca-Cola bottling in 1909. Baltzer was Secretary-Treasurer in Covington and remained there to operate the plant.

Soon a new bottling plant was needed, and it was constructed near the railroad tracks in the downtown area. This plant served the bottling needs until 1940 when Baltzer selected the third site for his bottling business; the location for a former cotton gin which had been destroyed by fire. An important asset for the new plant was the existence of a flowing artesian well to provide pure water for the bottling. The site also offered great visibility near the center of the city and along busy U.S. Highway 51.

Construction began in 1940, and on Sunday December 6, 1941 the final relocation to the new plant was underway when a radio broadcast interrupted the move to tell of the Japanese bombing of Pearl Harbor. The plant's opening always will be remembered, the next day, December 7, 1941, "Pearl Harbor Day".

Leander Baltzer acquired total ownership of the Coca-Cola franchise, and bottling continued at the Covington plant until 1979. In the later years his sons, Charles and George became involved in the business.

True to the Coca-Cola tradition, the Baltzers were actively involved in their community. The company maintained two sound trucks, which were used to announce high school football games, and provide public address services for a variety of community events.

Christmas was a special time for the Coca-Cola bottler as area residents enjoyed large holiday displays which were erected in city park in the 1930's. When the new plant opened, Christmas time included elaborate outdoor decorations and a second floor meeting room which

Coca-Cola float for homecoming parade

became a holiday wonderland. There was a large model railroad display, visits with Santa, and gifts for the children.

The plant's curved glass brick front wall provided easy viewing of the bottling process, and visitors were welcome to come inside where they would be treated to complimentary Coca-Cola as they watched the bottles being filled.

The Coca-Cola Company purchased and closed the Covington plant in 1979, with bottling being moved to the plant in Brownsville, Tennessee. Today the entire area is served by the CCR bottling plant in Memphis.

A tire company moved into the unique building in 1984, followed by a carpet business which occupied only a portion of the structure.

The old plant remained empty and decaying for almost ten years until 1996 when Rose Construction bought the building and began the task of historical restoration. Allan and Tim Rose were encouraged to take on the project by an area preservationist, Tim Sloan, and they received additional support from Middle Tennessee State University.

Much of the plant's original glass brick walls now curve through the first floor and lead visitors into a Coca-Cola museum area. Many of the museum items were located and purchased by Rose Construction. Others have been donated or loaned to the museum. Included are old Coca-Cola vending machines, trays, and display cases filled with memorabilia.

Rose Construction CFO/Treasurer Joe Griggs in the company's Coca-Cola Museum

As a reminder of the D-Day significance and World War II era of the plant, a framed set of twenty prints depicting wartime aircraft, are featured on a wall display near the entrance. The prints are a series produced by Coca-Cola during the war.

The original artesian well was still active and causing basement flooding when the restoration began. A method to drain the flowing water from the building had to be devised as an early step in the project. Much of the original interior construction has been preserved, and in doing the work an unusual hole was discovered on the second floor. Research revealed the hole was positioned on the floor of the second floor lab, and was

used to allow a container to be lowered to the bottling area below. A sample then would be obtained and raised back to the lab for quality testing.

The plant restoration also pre-served a large roof top Coca-Cola sign. Originally the signed included chaser type lighting, which was not working when the building was purchased. Attempts to have the Coca-Cola com-pany restore the chaser lighting have been unsuccessful. But Rose Construction now illuminates the sign at night with flood lights, and also have repaired and repainted the sign several times.

The unique plant is located at 126 U.S. Highway 51 South in Covington. The museum is open to the public 8:00 am to 4:30 pm, Monday thru Friday.

Chapter 18
Philadelphia, MS

When the Philadelphia, Mississippi Coca-Cola bottling plant was sold and ultimately closed in 1985, it was described as one of the area's oldest businesses. Now the old plant begins a new role in history as part of a country music museum, honoring Marty Stuart, Philadelphia's own nationally famous entertainer.

The building was constructed in 1926, but beverage bottling in Philadelphia goes back to 1906. At that time O.B. Fox, the publisher of the local newspaper, "The Neshoba Democrat", decided to get into the bottling business as well. For a brief period of time Fox conducted both operations in the same building.

Fox sold his interest in the newspaper in 1911, and eleven years later he purchased another bottling company, which had been owned and operated by two brothers, H. B. and Andrew J. Hutchinson. The continued growth of the bottling business required more space and in 1926 Fox constructed a two story brick building on Center Avenue. As

O.B. Fox, owner, Dr. Gully, Stanley Smith and Albert Jones

an ironic return to its' roots, the newspaper and the bottling business once again shared the building from the early 1930's to the late 1950's.

Coca-Cola actually was not bottled in Philadelphia until 1932, having been brought in by rail prior to that time. The bottling of Coca-Cola required more expansion at the plant, and in 1937 new equipment was installed which would fill 32 bottles per minute. Another equipment up-date in 1956 allowed the plant to increase production to 50 bottles per minute.

O. B. Fox died May 6, 1941, and the business was continued by his family. Fox's grandson Kenneth Lewis worked in the plant most of his life, and remembers at an early age standing on a box so he could wash bottles for his grandfather. Kenneth became manager in 1954 when he purchased the plant from his aunt Ava Johnson.

When Philadelphia Coca-Cola Bottling Company was sold in 1985, it was said to be one of the smallest bottlers in Mississippi with a staff of only 16 employees. The plant and territory became part of Northeast Mississippi Coca-Cola Bottling of Starkville. Harold Clark is President of Northeast, and his family has a long history of bottling Coca-Cola. His grandfather, C.C. Clark, was instrumental in the founding of plants in Corinth and New Albany in the early 1900's. Clark went on to expand Coca-Cola operations into other areas in Mississippi, as well as Kentucky and Tennessee. Today C.C. Clark, Inc. is owned by third and fourth family generations, conducting beverage business in five states.

Now the new life for the old Coca-Cola plant is under the direction of Community Development Partnership President David Vowell. However, for a while the "new life" for the plant was uncertain. The building had housed a furniture store until it was purchased by the county for storage in 2003. There were discussions in 2009 of demolishing the building to make space for a public parking lot. However, those plans were put on hold a year later when a professional assessment group called

David Vowell inspects old Coca-Cola plant as museum is planned

the building "an industrial landmark", and urged it be saved.

The Coca-Cola building was transferred to the development authority in 2013, along with state funding of $1 million for the project. The old plant is becoming an event center to host country music and other activities. An opening of late 2016 was planned.

David Vowell explained the center also will become a "warehouse home" for Marty Stuart's huge collection of country music memorabilia. Included are guitars, clothing, personal items, music and literature from famous entertainers including Johnny Cash, George Jones, Patsy Cline, Hank Williams and others. A separate country music museum is planned to provide a public show place for many of the items in Stuart's collection.

The state of Mississippi Country Music Commission is responsible for the state's popular "Country Music Trail". It has called Marty Stuart's collection "a living history of country music". The center was planned to serve as a combination of a theater, classroom, and history collection storage facility.

Marty Stuart was born in Philadelphia in 1958. A five time Grammy award winner, he has been recognized for over 40 years for his talent as a multi-instrumentalist, singer, song writer, photographer and historian. His talent was apparent at an early age, and when he was 13 years old he was touring with the legendary Lester Flatts. Stuart has teamed with many musical greats including Johnny Cash, Bill Monroe, Jerry Lee Lewis and Carl Perkins. His solo career exploded in the 1980's, as he earned platinum records and Grammys.

He told David Vowell now his desire in life is to share his memories in his Mississippi home town. A large historical marker honoring Stuart is located in downtown Philadelphia. The plaque was erected as part of the Mississippi Country Music Trail.

Coca-Cola memorabilia, including items from the Philadelphia plant, can be seen at the Beacon Street Antique Mall, 442 Beacon Street. Mall owner Clara Sims purchased many of the items from former plant owner Ken Lewis. Included are special recognition plagues which had been awarded to the plant during its years of operation.

Mall Owner Clara Sims with Coca-Cola memorabilia

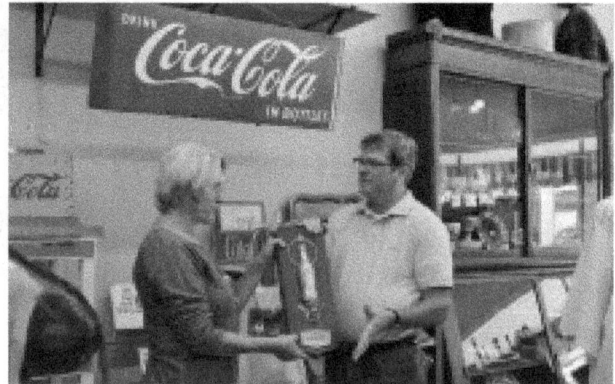

Former Coca-Cola plant owner Ken Lewis views one of awards presented to plant.

The new museum was dedicated July 9, 2016 in an event in which Marty Stuart participated. He unveiled an exhibit featuring his early years, called "Mississippi Boy. Marty Stuart the Neshoba County Years".

The exhibit has a special room in the museum and includes numerous personal items, as well as a video with Stuart describing his early days in Neshoba County.

Chapter 19
Morrilton, AR

A building built to become the fifth Coca-Cola bottling plant in Arkansas, has since served in other unique capacities in the city of Morrilton, but in 2016 the structure was facing an uncertain future.

The plant was built in 1929 by Frank Lyon, who also owned Coca-Cola bottling in Little Rock. A local news article described the building as both "colonial revival and modern movement" in style, with the traditional Coca-Cola signage. The plant's opening was an important event for the community, and was celebrated with dance music provided by a band from Memphis, and unlimited free Coca-Cola. The local newspaper chronicled the occasion with a 52 page special edition.

The building took on an exciting new role in the community in the 1960's, after Coca-Cola moved its operations to a larger facility. The former Crestliner boat plant on the east side of town had become available and met the growing needs of the Coca-Cola bottler.

Sam Walton, in the early stages of creating his future retail empire, purchased the former plant building and it became the nation's eighth Wal Mart store. It is said in the community that Walton approached a local business acquaintance seeking a partner for his new Wal Mart business idea. The friend, Wilbur Britt owned Ben Franklin stores in Morrilton and other Arkansas communities, and it was said that Walton offered him a partnership for $25,000. However, Britt apparently decided the proposed new Wal Mart business was too risky, and turned down the investment opportunity.

The old Coca-Cola building became a new focal point for Morrilton in 1978 after Wal Mart moved to a new location and donated the structure to the city. Mayor T. H. Hickey accepted the building, and for the next 38 years the old plant became home for the city hall and Morrilton police department.

However in 2016 the city acquired a former bank building and began making plans to abandon the Coca-Cola building. Mayor Allen Lipsmeyer expressed hope that someone would see the building's potential and create a fourth "life" for the structure, which is listed on the national register of historic places.

Ken Tucker became manager of Morrilton Coca-Cola after the move to the former boat plant. Ken said he learned the bottling business while working at the first plant under the guidance of Bob Coffman, who had been a Coca-Cola employee for 47 years.

Coca-Cola Morrilton became part of corporate Coca-Cola Refreshments when it was sold by Frank Lyon Jr. in 1985. Tucker remembers that soon after the sale, bottling was discontinued and the building was used as a warehouse and distribution facility for product produced in Little Rock. The Morrilton Coca-Cola facility finally was closed in July 2010.

However in 2016 plans were underway to convert that second Coca-Cola plant into a new attraction for Morrilton. Local businessman Mike Miller had purchased the structure, and work was underway for a local winery and craft brewery to occupy the building. Miller also saw the need for a large meeting facility in Morrilton, and his plans include a banquet hall to accommodate up to 400 people. A 4' x 8' Coca-Cola sign, which remained in the building, will become a featured display in the tasting room and gift shop area.

Chapter 20
Nashville, AR

To celebrate one hundred years of bottling and selling Coca-Cola in Nashville, Arkansas the local Coca-Cola company in 2011, created a display of memorabilia at the plant which has grown to become a permanent museum open to the public. Nashville bottling is the third oldest industry in the southwest Arkansas city, having been operated through the years by four generations of the Wilson family.

Kenneth Wilson - Pres. and sister Elizabeth Overton - Sec./Treas.

Soda bottling first began in Nashville at a grocery store owned by J. H. Moore. The store's back room contained a one-man, single bottle filler, producing soda flavors such as Nugrape, orange, lemon crush, and a secret formula drink called "Hot Shot".

W. W. Wilson and his son Forrest bought the store, but Moore continued the bottling for the Wilsons for a few years. Exciting news about a new drink called Coca-Cola came to the Wilsons' attention in 1910 and the following year they obtained a contract to bottle the unique new beverage. The agreement to bottle in Nashville was made with Walter Bellingrath from Mobile, Alabama who earlier had purchased the Coca-Cola rights to almost the entire state of Arkansas and also was bottling in Mobile and other locations.

The grocery store and bottling operation was moved to a larger facility in 1914, and the first accurate sales records show 315 cases of Coca-Cola were sold in the peak month of August that year.

Coca-Cola bottling continued to expand in Arkansas at that time and a new plant was established in Mena, about eighty miles northwest of Nashville. That franchise again was purchased from Bellingrath by Mena businessman W. G. Powers. Ultimately the Mena operation would be purchased by the Wilsons in 1950 and consolidated with the Nashville plant.

In the 1920's the Wilsons were able to complete the actual purchase of the Nashville franchise, and in 1924 acquired the De Queen territory from the Biedenharn family. De Queen is about 35 miles west of Nashville, but travel at that time was by dirt road. Once the highway was improved, De Queen Coca-Cola also was merged with Nashville.

Plant in 1937 with Forrest Wilson and 14 year old son Ramon on left side of photo.

Through the years Nashville Coca-Cola has been deeply involved in community projects. In 1939 Forrest Wilson purchased and then donated to the city the land for a baseball stadium. In addition he helped raise funds to build the stadium, which remains in use today by the high school and is called Wilson Park. Several state tournaments are held at the stadium.

Forrest also managed the town's semi-pro baseball team, which had as one of its opponents a team from Minden, Louisiana which was managed by Larry Hunter of Minden Coca-Cola bottling.

Nashville Coca-Cola often has been described as a "pillar of the community" as the company provides scholarships, assists with city park and other municipal projects, and has made possible the annual "Stand Up for America" July 4th celebration. Among those who have participated in the event were Glen Campbell and former President Bill Clinton, who played his saxophone.

Forrest Wilson's son, Ramon became president of Nashville Coca-Cola in 1944, and in 1981 witnessed the business expand to a new $2 million dollar plant at its current location. By 1985 sales reached a high of 650,000 cases.

Kenneth Wilson, the fourth generation to guide the business, became president in 1988. Also that year the plant was granted the area franchise for Dr. Pepper, and during the following years the Nashville operation consistently was honored as being the highest "per capita bottler" for Dr. Pepper in the United States.

Like most community bottlers, the Nashville facility stopped bottling on site in 2007 to concentrate total emphasis on product distribution and service.

An interesting chapter in the history of Nashville Coca-Cola has a "Bonnie and Clyde" connection. It was 1938 when the outlaws Floyd Hamilton and Ted Walters, who were alleged to be affiliated with Bonnie and Clyde, paid a surprise visit to the office of Nashville Coca-Cola. Apparently the two had traveled to Chicago to purchase firearms from the Al Capone gang. On their way back south they decided to rob a few banks along the way to pay for their recent purchases. They hit banks in Illinois and Memphis, and were working their way thru Arkansas when they decided to pay a visit to the First National Bank in Nashville.

However, local authorities had been tipped to the plan, and police were waiting on the bank's roof when the duo arrived in town. Needing a new target, the robbers went to the nearby Coca-Cola plant, where they entered the office with guns drawn, demanding "we need your money and need it quick".

Forrest Wilson removed the cash box from the safe. However it contained only some checks and $64 in cash, as the Coca-Cola trucks had not yet returned from the daily deliveries. Forrest convinced the robbers the checks would be of no value to them; so the famous "Coca-Cola hold-up" netted the robbers $64 for their efforts.

Sheriff's departments in neighboring counties were alerted, and a road block and shootout in Ogden failed to stop the pair. Ultimately the fugitives made their way to Texarkana where they hopped on a train and escaped to Dallas, only to be captured later and sent to Alcatraz prison.

It is said that Hamilton was one of the few inmates to ever escape over the walls at Alcatraz. He stayed on the rocks for three days, but hungry and cold, finally turned himself in and finished his term in 1963, the year Alcatraz was closed.

Hamilton became part of Nashville Coca-Cola history again in the 1970's. Ramon Wilson had learned the ex-holdup man had turned his life around and had become a lay minister in Dallas. During a subsequent trip to Dallas, Ramon located Hamilton, and the Coca-Cola robbery story was retold with some new details.

Hamilton's partner in crime was not as lucky. After being released from prison Walters also returned to Dallas, but was killed by police during another robbery.

Plans for expansion continue at the Nashville museum. Kenneth Wilson hopes to move the memorabilia to a larger space to accommodate more larger items such as vending machines, and even a 1926 Model T Roadster his father had purchased for fifty dollars. The car had been stored in a barn and was discovered by Ramon while operating the plant in Mena.

Ken Wilson, Nashville Coca-Cola President and his sister Elizabeth Overton, Secretary-Treasurer, planning to expand their Coca-Cola museum.

The museum is open to the public during business hours at Nashville Coca-Cola, 1301 S. 4th St.

Chapter 21
Indianapolis, IN

It once was called "the largest Coca-Cola plant in the world" as it produced over two million bottles of Coca-Cola per week in downtown Indianapolis. But in 2017, the massive 11 acre site was being prepared for an important new "life" in a major redevelopment project.

Built in 1931, a public grand opening celebration on September 29 attracted hundreds of visitors as well as Coca-Cola corporate officials including President Robert Woodruff and Vice President Harrison James. Also present was George Hunter of Chattanooga Coca-Cola, the nephew of bottling pioneer Ben Thomas.

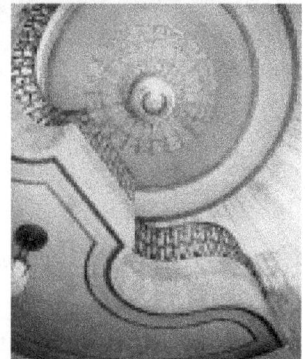

The two story building with 165 feet fronting on Massachusetts Avenue, featured a terra cotta facade, impressive indoor areas throughout the building, and elaborate designs of flowers, fountains and sun rays.

A garage building with similar architecture was located across the street. The plant was designed by Rubush & Hunter, one of the city's top architectural firms at that time. The firm also was involved in design work for nineteen other Coca-Cola plants throughout Indiana and Kentucky.

Bass Photo Co. Collection, Indiana Historical Society: Martin Collection, Indiana Historical Society

The plant's owner, James Yuneker had founded Yuneker Bottling Works in 1906. His brother L. E. Yuneker joined the company a year later at their first bottling plant at 710 E. Michigan Avenue. Bottling success came quickly and a new brick building was constructed at the current plant site on Massachusetts Avenue, with large additions being made in 1918 and again in 1930.

That first addition probably was necessitated by Yuneker's 1915 purchase of the city's original Coca-Cola bottling works, with the plant equipment being relocated to Yuneker's site. However both companies were operated as separate corporations until 1929 when they were merged to create Coca-Cola Bottling Company. Other soft drinks also were being bottled at the plant.

Soon after his 1915 Indianapolis Coca-Cola purchase, Yuneker also acquired plants in Gary, South Bend and Logansport. Sub plants were set up in 21 other Indiana communities.

Indianapolis Coca-Cola bottling had been sold by James Yuneker in 1965 to Anton "Tony" Hulman Jr. Hulman also purchased plants in Lafayette, Anderson, New Castle and Fort Wayne.

Hulman, a businessman from Terre Haute, always will best be remembered for purchasing the Indianapolis Speedway in 1945, and bringing the track back to become the world's most famous race course.

Hulman died in 1977, and four years later the family sold Indianapolis Coca-Cola to the Spectrum Group of Los Angeles, which had other plants on the west coast.

However the ink had barely dried on the sales agreement, when less than a week later Spectrum sold almost all of its' interest in Indianapolis to Houston businessman Marvin Herb, who had previous bottling experience in Indianapolis. He had been president of that city's Pepsi-Cola bottling from 1972 to 1976, when it was owned by Borden Inc. Herb had left Indianapolis in 1976 when he was promoted by Borden, and ultimately became an Executive Vice President of the company.

Herb said he purchased Coca-Cola for the opportunity to have his own soft drink bottling business. Two other top managers at Indianapolis Pepsi-Cola resigned to join Herb at Coca-Cola. At that time competition between the two bottlers was very close in central Indiana, with Pepsi-Cola having a slight lead. Herb said he planned to reverse that trend.

Later Herb acquired Coca-Cola plants in Chicago, Milwaukee and Rochester, New York. In 2001 he sold it all to Coca-Cola Enterprises for $1.4 billion.

Bottling continued at the Massachusetts Avenue location until 1968 when a new plant was constructed in suburban Speedway, Indiana. The abandoned 285,000 square foot building was purchased for $700,000 by the Indianapolis public school system, and it was used as a bus garage, warehouse and central kitchen.

A private developer, Hendricks Commercial Properties of Beloit, Wisconsin acquired the former plant building in 2017, beating out four other groups competing to redevelop the site. Hendrick's winning plan called for a $260 million project to include a hotel, apartments, a movie theater, offices and retail space. It was the costliest proposal of those submitted by the five bidders. It projected over $1 million alone to be spent on site preservation.

The plans called for saving a large portion of the original bottling building, including the exterior facades. It was believed the unique terra cotta remained in good condition, including the friezes of Coca-Cola fountains above the doorways.

Said to be one of the most architectural gems in Indianapolis, the site development proposals were to be approved by the Indiana Historic Preservation Commission.

The project developer, the Hendricks firm, was founded in 1982 and has been involved in five other major projects in Indianapolis. It was estimated the Coca-Cola redevelopment will provide a $5 million impact on the city's tax base, and create new resources for an important corridor in Indianapolis.

Chapter 22
Coca-Cola Ghost Signs

It all started in 2011 with the restoration of a faded Coca-Cola outdoor sign in Concord, North Carolina. The sign dated back to the 1960's. It was in bad condition and partially covered. Mayor Scott Padgett wanted it restored as part of the city's revitalization program.

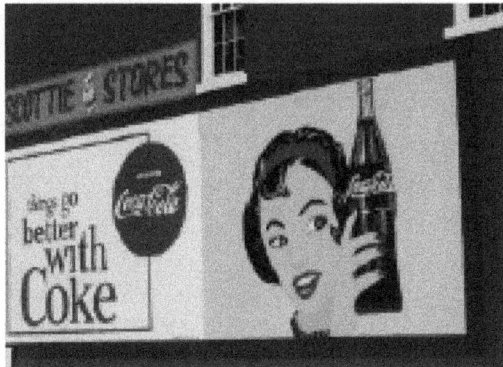

The mayor took his request to Coca-Cola Bottling Company Consolidated of Charlotte, and after some prodding, the nation's largest independent Coca-Cola bottler decided to get involved in the Concord project. That decision was the catalyst which convinced the company to restore over forty outdoor signs in just five years, and to make a commitment to continue the "ghost sign project". That first sign in Concord, when restored attracted considerable media and public attention, and became a landmark in the renewed downtown area.

The ghost sign project was coordinated by now retired Senior Vice President of Corporate Affairs, Lauren Steele. He realized the old signs often are an important part of the history of each town; a treasured connection to the past.

One of the largest refurbished murals is 17-feet high and over 60-feet wide located in Hinton, West Virginia. Originally painted in 1914, the sign has become part of another major downtown revitalization. Hilton boomed at the turn of the century as an important railroad and river town, but later the area suffered from a major decline in the economy. Now the Coca-Cola sign has an important community role adding to the vision of a new Hinton.

"The Ghost Sign Team" - Lauren Steele, retired Vice President Corporate Affairs; Alison Patient, VIce President Government Affairs; Emilie Nicholls, Communications Manager; Bob Bedell, Sr. Director Government Affairs

Before

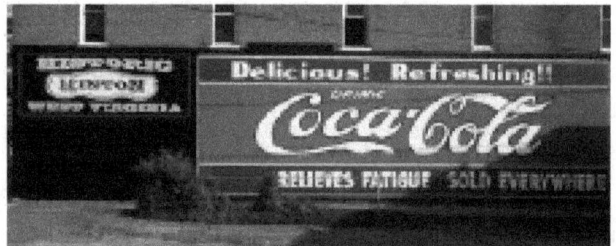

After

Another large restored ghost sign is a two story mural brought back to life in Hendersonville, North Carolina. The "re-birth" began after Mark Ray, the owner of a downtown building, discovered an old photo showing the mural on the side of his of his business. Ray is a self proclaimed history buff, and his business "Dad's Collectibles", includes the sale of Coca-Cola memorabilia.

Coca-Cola Consolidated responded to Ray's request for restoration, and the result was two repainted murals, one on each side of his building. Ray said his customers say the murals are "like a breath of fresh air". He added, "they are not billboards, they are historic."

The age of a ghost sign often is determined by the slogan painted at the bottom. One of the first was "Relieves Fatigue", others include: "Work Refreshed", "The Pause That Refreshes" and "You can Trust Its Quality".

Media attention created by a restored sign, will often result in another restoration project. Such was the case in Salisbury, North Carolina when the ghost signs project caught the attention of a group of young men who had a passion to revitalize their community's downtown area. They launched a Facebook campaign to gather support for restoring ghost signs in Salisbury. Coca-Cola Consolidated quickly got involved and moved in to refurbish two badly faded Coca-Cola signs.

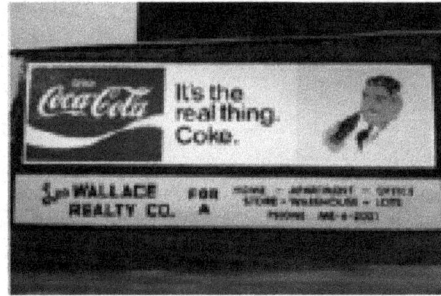

Before   After

Andy Griffith would have been proud to see a piece of "Mayberry" come back to life in June 2015 when the citizens of Mount Airy, North Carolina celebrated a restored downtown Coca-Cola sign. For many Mount Airy is better known as "Mayberry R.F.D.", as the community provided the location for the Andy Griffith TV show from 1960 to 1968.

A large crowd was on hand for the sign's ribbon cutting ceremony. Among those attending were Mayor Steve Yokeley and Barbara McMillian, the owner of the building where the sign is located.

Interest in the old sign got started in the 1980's after a building was torn down to reveal the 100 year old mural. A local resident and restoration enthusiast, Susan Ashley lead the efforts to get the sign restored, but she died in 1998 before being able to see her dream come true.

Building owner McMillian spoke during the ceremony and praised Ashley for her work saying, "we've waited a long time for this. It is one of the most beautiful pieces of art we've seen in downtown." The alleyway where the sign is located was renovated into a community event area, providing a location for concerts, outdoor dinning and other events.

Mayor Yokeley praised the role of Coca-Cola in making the restoration possible. Coca-Cola Consolidated provided Cokes and hot dogs for the celebration.

The first bottling plant in Mount Airy began operation in 1898. Coca-Cola Bottling Company of Mount Airy was started in 1917 by W. A. Jackson and B. F. Herman. It became part of Coca-Cola Consolidated in 1974.

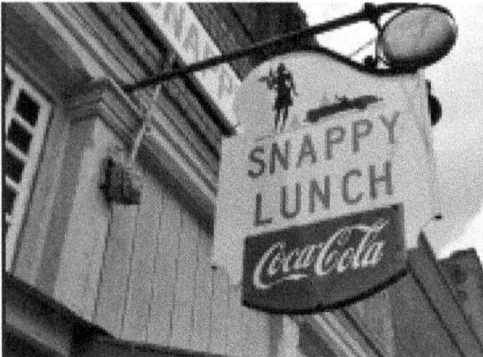

Another famous Coca-Cola sign also can be seen on Mount Airy's main street. It is the sign at Snappy Lunch, the popular Mayberry restaurant where Andy and Barney often enjoyed the popular pork chop sandwich.

The Johnson City, Tennessee historic district redevelopment has benefited by the restoration of downtown Coca-Cola signs. The first restored sign is located across the street from the remodeled historic Tweetsie Railroad Depot. The East Tennessee and Western North Carolina Railroad, called "Tweetsie" constructed the depot in 1891. The property once considered for condemnation, in 2016 had become home to the Yee-Haw Brewing Company and the Duck Taco Shop.

The unveiling ceremony for the first Coca-Cola mural took place in July 2015 at the nearby Founders Park. Guests for the event enjoyed a catered lunch and the opportunity to personalize a mini-Coke can with their name at the "Share A Coke" display.

A second Coca-Cola sign in the historic district was restored a few month later. The downtown development manager praised the restoration saying the signs help bring new attention to the historic area.

When the two signs were completed they were not immediately covered with a protective clear coat finish, which allowed the signs to fade and weather for a time, so they would better fit with the historical integrity of the area. After aging the final finish was added to the signs. Sign zoning regulations are in effect in the district.

Creating outdoor Coca-Cola signs for over fifty years allows the unique talent of sign painter Andy Thompson to be enjoyed by hundreds of thousands of people for many years to come. In 2017 at the age of 73, Andy was working his brush stroke magic for the ghost sign project of Coca-Cola Consolidated.

Thompson started painting signs for Coca-Cola Consolidated in 1958, shortly after graduating from high school in Charlotte. He retired after thirty years when outdoor signs were no longer being hand painted. But his brushes did not get packed away, as the bottling company kept him busy again as an independent contractor, working on sign projects throughout their territory.

The ghost signs project became Andy's new mission after he completed the first one in 2011 in Concord. That initial restoration was so well received that it established a new sign painting direction for both Andy and Coca-Cola Consolidated. Since then Andy has done almost all of the ghost sign restorations, and after more than a half century, he says he has painted more Coca-Cola signs than he can count. But he does have special memories of the enthusiasm in the communities where the ghost signs are brought back to life. The growing list of requests for Andy to work his sign magic should keep him busy for many years.

When Coca-Cola Consolidated needs the help of a second sign painter, they call on Jack Fralin of Roanoke. He has done a restoration in his hometown as well as another in Hendersonville. Jack also has been called upon by Coca-Cola Corporation, and in 2016 did an ouside wall mural at the popular Manuels Tavern in Atlanta.

Coca-Cola has been part of American life for over 130 years, The ghost sign restoration reflects the enduring connection between Coca-Cola and the American Experience.

Chapter 23
Griffin, GA

Coca-Cola memories inspired a Georgia businessman to save a 75 year old former Coca-Cola bottling plant, and to create a "new life" for the building.

John Carlisle of Griffin, Georgia first visited the Coca-Cola plant at 410 E. Taylor Street when he was participating in a Boy Scout tour of the facility in 1953. He has other Coca-Cola memories as well, including those experiences his father,

Ernest Carlisle would tell about working at a local drug store. One of Ernest's duties at the store was to serve as a soda jerk, where he would prepare and serve Coca-Cola fountain drinks. He also would explain to young John how the Coca-Cola syrup, in those days, arrived in large barrels.

Coca-Cola was first bottled in Griffin in 1907, in a small plant managed by
J.C. Vaughn. H.B. Montgomery assumed
management of the plant in 1912, a position
he would hold for over 40 years. The plant
was moved to the Taylor Street location in
the early 1940's and remained there until it
was closed in 1956. The last plant manager
was A. Talley Montgomery.

John Carlisle saw the old building as a landmark, which he said was not
being appreciated. He believed it needed to be saved, not only as an important
historical structure, but also as an important future place within the community.

He decided he would try to purchase the old Coca-Cola building, even
though a previous owner had failed in an attempt to renovate the structure. At
that time the ownership was being held by the FDIC which had provided financial
backing for the failed project.

John bought the 42,000 square foot building in 2011, and within two years
created a new location for over 30 Griffin businesses. Included were a restaurant,

clothing stores, a yoga studio and health facility, travel and insurance companies, an ambulance service, and a variety of professional offices.

John explained, "People have a positive feeling about the building."

He also sees his building being used for community events, and actually created an attraction in 2013. Residents were encouraged to provide a photo of their own Coca-Cola memory. The photos featured people with Coca-Cola as part of their everyday life. They became part of a large photography exhibit inside the building which John called "The Feel Good Emblem of Our Society".

Another Coca-Cola connection in Griffin finds the great granddaughter of Asa Candler as a resident of the city. Mettelen Moore was a prominent member of the community from 1961 until she died in July of 2014.

Her mother, Asa's granddaughter, was Lucy Candler Thompson. Her father, Thomas Homer Thompson, owned a Coca-Cola bottling plant in Galveston, Texas when she was a very young girl.

Mattelen and her husband Jared Leon Moore moved to Griffin in 1961. However, neither were involved in the Coca-Cola business. He was a hospital administrator, and she was a school science teacher, who also was very active in several community arts organizations. After her death a newspaper memorial honored Mettelen as a "patron, friend, and supporter of local arts organizations for over thirty years".

Her daughter, Lucy Cawthon, son Jere Moore, and their families reside in the Griffin area.

If you visit the town of Griffin a stop at Gigi's antiques, 113 E. Taylor, will often result in the opportunity to shop some interesting Coca-Cola memorabilia displays.

Chapter 24
Monahans, TX

A Coca-Cola museum, which is part of an unique west Texas historical attraction, helps reveal the story of Coca-Cola bottling in the Monahans, Texas area.

The Coca-Cola museum building is featured at the unusual site known as "The Million Barrel Museum". The barrel actually is a giant earthen bowl, which had been dug 35 feet deep and paved with cement to hold excess crude oil being pumped during the west Texas oil boom in 1928.

Bottling in the area began in 1925 when John Dunagan, a traveling provisions salesman from Midland, built his first plant in McCamey. However a year later he relocated to a facility in Midland, where he was able to obtain Coca-Cola sub-bottling rights from Hope Smith of El Paso.

A year later a fire at his plant in Midland damaged the machinery and destroyed most of the inventory. Again John moved to a new location. This time to neighboring Monahans where the water quality was described as excellent for bottling, and the oil boom was underway.

John had two sons, Conrad and Robert Louis. Conrad, who was the eldest, began working in the Monahans plant in 1932, when he was just 17 years old. A bottling industry publication at that time reported Conrad "probably was the youngest bottling plant manager in the country". A few years later Robert also became involved in the plant. Also the boys' grandfather, C. B. Dunagan had some duties there as he answered the phone and handled financial matters.

The town of Monahans continued to experience rapid oil boom growth, and a new Coca-Cola plant was constructed in 1938.

The Dunagans acquired another Coca-Cola plant in Marfa, about 125 miles southwest of Monahans, in 1957. Robert became the sole owner of the Marfa facility in 1961 when he and Conrad decided to divide their territory, with Robert receiving the southern division. Robert then relocated to Alpine, about 25 miles east, where he built a new plant, Big Bend Coca-Cola, to replace the bottling being done in Marfa. Robert's three sons, Patrick Michael, Robert and Dan became involved in the Alpine plant in the 1970's.

Photos: Ward County Archives, Monahans

Meanwhile in Monahans the plant's success continued, and the company built its' third...and what would be final...plant in that city in 1950. During those last years of operation a gas leak resulted in an explosion in 1980 which resulted in considerable damage and injury to seven workers. The damaged area was repaired, but three years later Conrad retired and sold the plant to the Coca-Cola bottling company of Lubbock. Bottling in Monahans continued until 1986 when Southwestern Coca-Cola of Amarillo purchased the territory and closed the plant.

The Big Bend Coca-Cola plant in Alpine remained a Dunagan family operation until 1999 when it was sold to Coca-Cola Enterprises. When the family owned Big Bend, they also maintained warehouses in Pecos and Fort Stockton, as well as in that very first plant building in McCamey.

A visit to the Monahans Coca-Cola memorabilia display includes the added experience of the Million Barrel Museum. The huge in-ground oil tank was the idea of Shell Oil, trying to find some way to store the oil rapidly being pumped from the wells. However, all was lost when the crude oil weight of 315 million pounds caused the cement liner to crack, with almost all of the oil running back into the ground.

Thirty years later local entrepreneur Wayne Long, who owned a trailer park, envisioned the abandoned barrel as a future west Texas oasis with fishing, boating and swimming. He purchased the land with the giant earth dish, sealed the cement cracks, and filled it with water. But his dream was short lived, as the barrel was no better at holding water than it was oil.

Long later thought the high banked walls would make the barrel an ideal place for auto racing, but nothing developed from that idea.

When Long died his widow donated the land to the Ward County Historical Commission, which created the Million Barrel Museum in 1987. The Coca-Cola museum building is located there, along with other historical structures moved to the site, including an old hotel and Monahans first jail.

Many of the old Coca-Cola items on display came from a Monahans couple who had ties to the old Coca-Cola plant. Dan Wetzig had been bookkeeper and paymaster for over ten years at the plant. Dan and his wife decided to open a restaurant after Dan left the plant. They operated the Big Burger restaurant for over 35 years, all the time collecting Coca-Cola memorabilia and antiques, and creating a local landmark which attracted visitors from throughout the country. Some of the collected items came from the days when Dan worked at the plant. Others were given to the Wetzigs by area residents who may have had them stored in an attic or shed, or were acquired during a trip.

The Wetzigs were forced to close their Big Burger in the spring of 2009 after Dan was diagnosed with a rare form of muscular dystrophy. In the months which followed, the community honored Dan and a special event was held which attracted more than 1000 people. One of those who helped organize the tribute was 95 year old Kitty Dunagan, the widow of Conrad.

After Dan's death, the large private collection was donated by Elaine to help create the Coca-Cola museum at the Million Barrel. Elaine said she wanted people to continue to enjoy the many items she and Dan had acquired through the years.

The museum is open Tuesday-Saturday, 10am to 6pm. Visitors are advised to call first to verify.

Chapter 25
Columbia, MO

An entertaining stop on the Coca-Cola Trail is Columbia, Missouri where you can enjoy a movie in a former Coca-Cola bottling plant.

The plant was built in 1935 by Ed Roberson, who had purchased the Coca-Cola Bottling Company four years earlier from C. R. McCallister. Roberson was from Newport, Arkansas and his father had operated Coca-Cola plants in Missouri and Arkansas.

Photo credit: Laura Lee Elfriet

Sketch credit: Laura Lee Elfriet

The old Columbia plant, located at 10 Hitt Street, was recognized for its unique Colonial Revival style, and is included on the National Register of Historic Places.

Sketch from the 1936 "Columbia Missourian" newspaper included in applications for Historic Places registration.

However, it was not the city's first bottling plant, as records indicate bottling was being done in Columbia as early as 1877. The Columbia Bottling Works began operating in 1917, and became Coca-Cola Bottling Company in 1925.

Bottling continued at the Hitt Street plant until 1966, when Roberson moved to a new location on south highway 63.

Photo credit: Laura Lee Elfriet

Ed Roberson died in 1975. Columbia Coca-Cola was purchased by Coca-Cola Mid-America of Kansas City. The franchise went through two other ownership changes before becoming part of Coca-Cola Refreshments in 2010.

The Hitt Street building became home in 2008, to Ragtag Cinema, a local film society which provides a venue for the best U.S. Independent and international films.

The society began as a volunteer organization in 1997, showing films in a downtown concert facility. It relocated in 2000 to a small cafe setting with seating for only about 70 people.

After Ragtag became a not-for-profit organization in 2004, a community fund raising campaign raised $250,000 to convert the Coca-Cola building into a cinema facility. Sharing the building with Ragtag are two other tenants, the Uprise Bakery and Hitt Records.

Photo credit: Laura Lee Elfriet

A highlight for Ragtag is an annual four day film festival.

Chapter 26
Charlottesville, VA

An historic Coca-Cola plant in Charlottesville, Virginia has become a new focal point of the community, as the renovated structure has become the home for several new tenants which include retailers, the University of Virginia, and a popular Alpine type beer hall and garden.

Brothers Joshua Hunt and John Woodruff created "Kardinal Hall, Beer Garden and Bocce". The attraction provides seating for 250 inside, 150 on an outside patio along with two outdoor bocce courts, ping pong and board games. A full menu and wide variety of beers are available.

The former bottling plant also is home to the University of Virginia licensing and ventures group, which provides assistance for new business startups in the community.

The first tenant to actually open in the plant was Timbercreek Market. The owners, Zach and Sara Miller, moved from selling vegetables at their farm stand to an expanded operation in their new

location. The market offers a cafe with a full lunch menu, a cheese shop and meat market.

Other tenants include Blue Ridge Cyclery, a complete bicycle sales, service and rental business; and The Juice Laundry, an organic juice and food retailer.

Building owner, Riverbend Development, purchased the 38,000 square foot structure in 2013. The building had been placed on the National Register of Historic Places because of its' historic value and art deco style.

Designed by Washington architect Doran S. Platt, the building is set off by a recess entry which includes an overhead 11-foot wide stone panel with script lettering "Coca-Cola Bottling Company 1939".

When renovating the building Riverbend owner Alan Taylor faced the challenge of preserving the historic features, with changes to the structure being approved by the Virginia Department of Historic Resources.

The plant, built in 1939, was like many designed at that time, to assure the public of clean and sanitary bottling conditions. A clear view of the bottling operations was an important aspect of the design. The plant was the third and last Coca-Cola bottling location in Charlottesville.

James E. Crass and his son-in-law, Walter Sams, opened the first plant in a rented building in 1920. Sams was born and raised in Georgia, while Crass was from Kentucky, and was one of the pioneer Coca-Cola bottlers, opening his first plant in 1904 in Richmond.

Crass had first become interested in the new Coca-Cola business opportunity when he met Ben Thomas while working in Chattanooga on the Lookout Mountain Incline Railway. He and Thomas both were from Mayfield, Kentucky and would talk when Thomas rode the railroad to his home.

That first Charlottesville plant, in a former agricultural implement ware-

house, quickly became busy, no doubt because of Crass's previous Coca-Cola bottling
experience. After only three months of operation, the plant was using Coca-Cola syrup at
an annual rate of 10,000 gallons.

Crass
and   Sams
constructed the second plant in 1926, a two story
brick structure at 132 10th street. That structure in
2016 was still standing and being used for rental
apartments.

Photo: Ryan Kelly, Daily Progress

Crass remained as general manager of the second plant, and C. S.
Hunter Jr was listed as manager. Gross sales in 1926 were $65,779.

It is interesting to note the contracted cost to purchase Coca-Cola syrup at
that time was $1.30 per gallon. The contract also contained a clause which
provided for an increase in the syrup price of six cents per gallon for every one
cent per pound increase in the cost of sugar.

The editor of the Coca-Cola Bottler magazine visited and new plant and
described its appearance later when he wrote, "this little plant is well lighted and
built in such a way as to be kept sanitary."

By 1930 the plant was selling over 134-thousand cases per year, but then
production decreased during the following Great Depression years.

As the economy improved production again increased and by 1936 the
Charlottesville plant was selling over 1.5 million cases per year. It was that rapid
growth which lead Sams to the decision to again build a new and larger facility, and
in 1938 land was purchased and the plans were made.

Crass did not live to see the construction of the third plant, as he died in
Richmond in 1930 at the age of 53. He had succeeded in creating one of the largest
Coca-Cola bottling companies, with a final total of 42 plants in five states. The
Crass Coca-Cola group was sold in 1978 to the Simplicity Pattern Company of New
York. In 2016 the former Crass plants were part of the large Coca-Cola Bottling
Company Consolidated of Charlotte, NC.

Bottling was stopped in 1973 in the Charlottesville plant, but the building
remained as a regional distribution center until being closed in 2010.

Chapter 27
Ocala, FL

An architecturally unique Coca-Cola plant in Ocala, Florida was closed in 1962. But the building, which was placed on the National Register of Historic places, continues to attract attention because of its impressive Mission/Spanish revival design.

Fort Lauderdale Plant

The plant was designed in 1939 by Courtney Stewart and his associate Alexander Martin of Fort Lauderdale. The Ocala plant incorporated much of the same design as the Fort Lauderdale Coca-Cola plant, which they had designed a year earlier.

Records indicate Coca-Cola first was bottled in Ocala in 1908 by the Ocala Bottling Company. The business was located in an old wooden structure built in the 1800's, which later housed a hotel, grocery, and a furniture store. The building last was home to Southern Plate Glass and Paint before being demolished in 2012.

Coca-Cola faced competition in those early years, as Chero-Cola bottling also was operating a plant in Ocala in 1915. Chero was a cherry flavored cola developed by a Georgia pharmacist in 1910. However, the term

"cola" was dropped from the beverage's name after a successful trademark infringement suit had been initiated by Coca-Cola. Later the Georgia

company created R.C. Cola and the Nehi line of beverages.

Coca-Cola bottling in Ocala was moved to a new plant, near the original site, in 1926. Frank Owen Hicken was transferred from Coca-Cola in Jacksonville to become manager, and held that position until the Ocala operation was closed.

Ocala's last Coca-Cola plant was built at a cost of $44,000. The building permit issued for the construction was the largest in the county since 1927, and was said to be a sign of the end of the depression years.

The Coca-Cola plants in Jacksonville, Ocala, Fort Lauderdale and other Florida communities were part of Associated Coca-Cola one of the largest bottlers in the nation. Headquartered in Daytona Beach, the bottling empire was owned by the Chapman Root family, also famous in Coca-Cola history for creating the iconic Coca-Cola bottle in 1916 at their bottle plant in Indiana.

The plant began operation in 1940 and soon would help meet the rapidly growing refreshment needs of a popular tourist attraction called "Silver Springs". Locals remember that in those "pre-Disney" days an increasing number of now traveling visitors were making the Ocala area Florida's fastest growing attraction.

Silver Springs was one of the largest artesian springs ever discovered, and featured their famous glass bottom boats. The attraction gained even more recognition as it became the location for the filming of six Tarzan movies featuring Johnny Weismiller.

After Coca-Cola left Ocala, the unique plant building was used for a variety of purposes, but fell out of public attention until it was bought in 2002 by Jack Gartner, owner of the Gartner Group, a local general contractor.

The new owner placed emphasis on converting the structure to its original design, as he created the Grande Pointe event center. The original bottling area became a large banquet hall to host wedding receptions and other functions. Additional meeting space was created on the first

floor and mezzanine area, and an elaborate conference room and offices occupied the upper level.

Coca-Cola memorabilia inside the building and a cast stone shield on the outside serve as a proud reminder of the history of this special building.

The Ocala area became part of Coca-Cola Beverages Florida (CCBF), which was formed in May 2015 through territory acquisition from the Coca-Cola Company. Headquartered in Tampa, CCBF is among the largest Coca-Cola bottlers in the nation.

Chapter 28
Hattiesburg, MS

An aging Coca-Cola bottling plant building in Hattiesburg, Mississippi was saved from ultimate collapse thanks to the efforts and vision of a local businessman who also has an interest in area history. Ken Dickinson's passion, when he purchased the building in 2004, was to bring it "back to life". But he may not have anticipated the challenges, which would include hurricane Katrina damage in the middle of his restoration project.

Original Hattiesburg Coca-Cola Plant

The building, located at 126 Mobile Street in the city's historic district, was the second location for Coca-Cola bottling in Hattiesburg. The first plant began operating in 1906 on 2nd street as Hattiesburg Coca-Cola Bottling Company, and was welcomed with a reported first

week total sales of $275.

The Thomson family owned and operated the company for 86 years. William Alexander Thomson managed the company for 46 of those years, from 1921 to 1967. William faced a major production challenge during the war time years of 1942 to 1945, to deliver Coca-Cola to meet the demands of nearby Camp Shelby. Operating on an overtime schedule, the little plant produced more than one and one-quarter million cases of Coca-Cola per year, which was more than twice the amount the plant had been designed to produce.

W.A. Thomson, pictured left, managed Hattiesburg Coca-Cola Bottling Company from 1921-1967.

William's son Richard S. Thomson began employment at the plant after returning from service in the U.S. Marines. He advanced to become president and CEO of the company, which experienced expansion and continued growth under his leadership.

Hattiesburg Coca-Cola's Current Facility

The Coca-Cola plants in Picayune and Columbia were consolidated to Hattiesburg, and in 1955 a new plant was built on highway 49 south. Growth continued and in 1978 another move was made to a larger facility at 201 Coca-Cola Avenue.

The new plant included large public meeting facilities which often served as an "unofficial convention center" until a city owned center was constructed in 1998. Hattiesburg Coca-Cola was sold in 1992 to become part of Coca-Cola Bottling United, headquartered in Birmingham.

Richard Thomson retired to Bay St. Louis on Mississippi's gulf coast, where he owned an elaborate home the locals referred to as "the Coca-Cola plantation". Richard died in Bay St. Louis in 2003 at the age of 77.

Meanwhile the old plant building on Mobile Street continued to deteriorate and was only being used for occasional storage by local businesses.

The roof was starting to collapse when Ken Dickinson first surveyed his new purchase. Rainwater had gotten inside and most of the first floor ceiling had fallen into a pile of debris on the floor of the former bottling area. Ken was able to

salvage some of the ceiling support beams for future use. An indoor elevator shaft, large enough to accommodate a vehicle, had to be demolished and removed. In its place a large staircase was built for access to the second floor, where the old plant offices had been located. The Hattiesburg plant workers made the company's own wooden Coca-Cola cases, and that construction, along with storage and employee lockers, was done on the upper level.

A painted glass Coca-Cola sign, which was over 100 years old, was salvaged from a front window.

Ken Dickinson in an old elevator shaft.

Soon after the large glass sign had been saved, hurricane Katrina struck Hattiesburg in 2005, and the remaining front windows were destroyed, along with heavy damage to the building and much of the restoration work already done, including damage to the new roof.

A full service commercial kitchen was built, and in 2006 the old plant opened as "The Bottling Company", a restaurant, bar and entertainment facility which was operated for four years by Dickinson's nephew. Next the old plant became known as "The Shed", operating under new management as a barbecue and blues restaurant.

Dickinson sold the restored building in 2012, and it was operated for a brief period as multi-purpose event center.

A new life for the old plant began in January 2016 when it was purchased by Ron Savell, who also owned several restaurants in Hattiesburg. Savell's plans for a high quality event facility became reality six months later with the re-opening of "The Bottling Company", as a popular wedding and private function location, under the management of Kelly Wagner. The century old etched glass sign and other Coca-Cola memories are seen throughout the building.

One of the best places to search for Coca-Cola memorabilia in the area is the Highway 49 Antique and Flea Market, which is owned and operated by Ken Dickinson, the

Ron Savell and Kelly Wagner review event plans in front of old Coca-Cola sign.

person who saved the old Coca-Cola plant. Located eight miles south of Hattiesburg, Dickinson's market features 200 vendors in an air conditioned building, as well as unique outside exhibits.

One of Ken's personal Coca-Cola treasures is a brand new 1920's Coca-Cola sign, which he discovered in an unopened box when renovating the Hattiesburg plant.

Ken displays the rare sign, along with other memorabilia at his market.

Chapter 29
Black Hills Gold and Coca-Cola Memories

The Black Hills of South Dakota may best be remembered for the gold mining days. Many memories of those days live on with old Coca-Cola outdoor signs in Lead and Deadwood. In fact the area may be home to one of the nation's oldest Coca-Cola murals.

South Dakota gold was first discovered at French Creek during the George Custer expedition of 1874. News of the discovery spread fast and within a couple years over 4,000 prospectors were seeking their fortunes in the streams throughout the area.

When those rich placer deposits were depleted, hard rock mining began in the 1880's. Research indicates the area's first outdoor Coca-Cola sign appeared just a few years later in 1897 in Lead. If the records are correct it may be one of the nation's oldest, as what has been recognized as the first sign was painted only three years earlier in Georgia. In 2017 discussions were underway to restore the sign, as two other Coca-Cola

murals, one in Lead and the other in Deadwood, already had been restored.

Profitable mines were located near the two towns, but at many mines the gold played out quickly and most were closed in the early 1900's. However, the Homestake mine in Lead proved to be the "mother lode", and the mine continued to operate for 126 years, finally closing in 2002. At one time gold mining was the area's most important industry.

Drilling Homestake gold on Main St.

Sign painter and local historian Les Rosales restored the Coca-Cola signs in Lead and Deadwood, and was anxious to restore the 1897 sign in Lead. The sign is located on the side of the Stamp Mill restaurant on Main street. The building was built as a general store in 1897, and historic photos indicate the sign was painted on the building that same year. Through the years the sign was protected from weather as the large Homestake opera house was built next door in 1914. The sign in Lead is original, not restored. Another original sign, believed to have been painted in 1907, was recently discovered in Opelika, Alabama.

Stamp Mill sign protected by nearby opera house wall. Photo credit: Richard Carlson

Sign painter Les Rosalles and Art Project chairperson Joan Irwin

Rosales was selected by the city's Urban Art Project to paint Lead's first Coca-Cola restoration in 2016. The large 50' x 60' sign is located on the senior center building, which was built in the 1920's as a hardware and auto parts store. The art project had obtained a matching grant of $3600 to make the restoration possible. Project Director Joan Irwin praised Rosales' work saying the Coca-Cola sign is a signifcant contribution to improving Lead's main street.

Rosales, a self taught sign painter, has been creating outdoor signs since 1963, but did his first Coca-Cola restoration in 1989 in Deadwood. That 48' x 20' sign is located on the historic Bodega building, which was built in 1877 and is one of the oldest establishments in the town. Originally known as the "Buffalo Bar", it was named after the original owner's friend and frequent customer, Buffalo Bill Cody.

Restoration underway

Through the years the building housed a variety of businesses, but in 1989 it got a new life when the city of Deadwood approved gambling, and a new owner, Bob Regan commissioned Rosales to restore the Coca-Cola mural. Now called "The Buffalo Bodega", it is a full service casino with gaming, dining, and entertainment.

Coca-Cola was first bottled in Lead in the 1920's by Richardson Ice Cream and Bottling Works, in a building located across from the Homestake mine. Richardson had acquired a franchise for territory west of the Missouri river. His Coca-Cola business was

sold in 1949 and was moved to Rapid City to become part of Coca-Cola Bottling Company High Country. The building in Lead continued as a Coca-Cola warehouse until 1955, when it was sold to Wayne and Bonnie King to become King's Grocery.

Artifacts and photos of Lead's first Coca-Cola plant can be seen at the Black Hills Mining Museum. The public museum is located just a few doors away from the Stamp Mill Restaurant, where visitors also can view the historic 1897 mural.

Chapter 30
Along the Trail

Grayling, Michigan

A local family restaurant has become a popular Coca-Cola museum thanks to the vision and efforts of a businessman in Graying, Michigan.

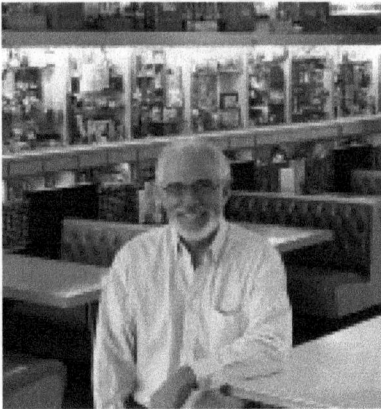

It all began in 1938 when Earl Dawson founded Dawson's Drug Store, which included a soda fountain. The business was destroyed by fire in 1957, but the soda fountain was saved and the business re-opened a year later and operated as Dawson's until 1994 when it was bought by Russell and Jane Stevens, and became Dawson & Stevens Diner.

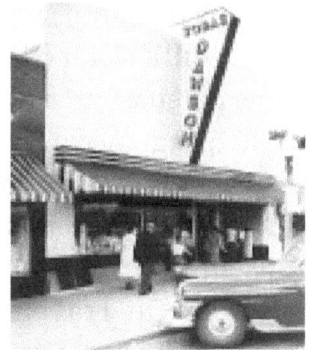

In 2003 radio station owner and local businessman Bill Gannon, looking for a new challenge, bought the downtown diner. For over a year Bill closely observed how things were being done, before closing the diner for an 18-month major remodeling project to create the "Dawson & Stevens Classic 50's Diner & Soda Fountain".

It was during the remodeling when Bill had the opportunity and vision to create a Coca-Cola museum in his new restaurant. By chance he learned of a place called "The Bottle Cap Museum", located about 25 miles away in Sparr, a tiny village near Gaylord. The museum owner, Bill Hicks had been collecting and displaying Coca-Cola memorabilia for about fifteen years, but called his place "The Bottle Cap Museum" because he had been told he could not call it The Coca-Cola Museum. However his exhibits did include a rare bottle cap collection featuring NFL players from the 1960's.

Bill made his first visit to the museum to purchase some old Coca-Cola bottles for his restaurant. Hicks responded by offering to sell his entire collection of 8,000 items, saying he wanted to retire and his desire was to sell everything to one person who would put it all on display.

Gannon balked at the price Hicks mentioned, but as he returned to Grayling he envisioned the collection in his restaurant. It took another visit and some serious negotiations with Hicks and his wife, before the Bottle Cap Museum would be headed for its new restaurant home. Bill already had installed 50's style neon and lots of stainless steel, and the Coca-Cola memorabilia would fit right in.

Bill's sister Marianne McEvoy, after some persuasion, agreed to take on the task of helping move the museum items to displays she would create at the restaurant. Marianne continues as curator of the museum and makes frequent trips from her home in Traverse City to update the Coca-Cola displays.

The museum collection brought with it an expanded mission for Gannon, as he became an enthusiastic Coca-Cola memorabilia collector. His on-going search has resulted in a personal collection of over 20,000 items which allows the diner to up-date displays as items are rotated in and out of storage. Bill only sells an item if it is a collection duplicate.

What may be a one of a kind item, is a Coca-Cola Indy style go-kart. The niece of the previous bottle cap museum owner had won the kart many years ago with a one dollar raffle ticket. Coca-Cola representatives have said they have never seen another one.

Another museum "treasure" is a restored 1950's Coca-Cola delivery truck.

Bill had located the truck which was abandoned in the woods in Minnesota, and had it shipped across Lake Michigan to become a museum attraction.

The unique diner museum has welcomed visitors from throughout the United States, as well as from other countries including Japan, New Zealand, and Russia.

Breakfast, lunch and dinner at the restaurant include menu items with 50's and 60's names.

Former bottle cap museum owner Bill Hicks visits the diner from time to time to see his collection and share a few stories. And a special visitor was an elderly gentleman with a young girl, who explained why they would sit in a specific booth. He said, "That is where I proposed to your grandmother."

New Orleans, Louisiana

The term "Swamp Pop" in Louisiana is used to identify a style of music played by Louisiana musicians, and often refers to older, traditional songs popular throughout the state. However, mural artists Robert Dafford of Lafayette, put a new twist on Swamp Pop when he created a mural in New Orleans.

Bottles of Coca-Cola are featured in a Louisiana swamp scene in the unique painting which covers a large lobby wall at the Coca-Cola plant in New Orleans. Plant management commissioned the work when their new facility was built two decades ago, and the colorful mural has provided a special Coca-Cola memory for visitors, especially youngsters enjoying a plant tour.

Sales Manager Charles Mader notes that New Orleans is the largest facility in the giant multi-state Coca-Cola United Company, handling over 23 million cases per year. United acquired the New Orleans territory in March 2016.

Mural artist Dafford also has painted Coca-Cola murals on river walls in Vicksburg and Paducah.

Los Angeles, California

"The Good Ship Coca-Cola"

A visit to Los Angeles provides the opportunity to view the only Coca-Cola plant in the world built in the shape of an ocean liner. Located at 1334 S. Central Avenue, the building was built in 1936 when Stanley Barbee was president of Los Angeles Coca-Cola.

Barbee was an avid yachtsman, and had the streamline Moderne style building designed by Robert V. Derrah, who earlier had designed a Hollywood shopping center which featured a ship shaped structure.

The Coca-Cola building has two rows of porthole windows, a ship's catwalk, and ship railings. The Coca-Cola sign on top was designed in the shape of a ship's bridge. Wooden mahogany floors and hand rails are used inside the building.

Coca-Cola bottling was moved to a nearby three story brick building, which in 2014 was converted to an elaborate office and retail complex. The ship style building continues to serve as offices for Los Angeles Coca-Cola, but only can be viewed from the outside as tours are not given.

It is said the ship design was first opposed by Coca-Cola officials until they were assured future additions could be made to the structure.

The building was designated a Los Angeles Historic Cultural Monument in 1975.

Kansas City, Missouri

One of the tallest structures built for Coca-Cola in the early 1900's was in Kansas City. Coca-Cola Company owner Asa Candler had the 12-story building constructed in 1914 at the corner of 21st and Grand, where, a year earlier, he had purchased the property near the city's new Union Station.

Coca-Cola's construction chief, Anthony Tufts of Baltimore, had designed Coca-Cola buildings in cities throughout North America, including New York, Chicago, Dallas, Baltimore and Atlanta. The Atlanta building, constructed in 1905, was 17 stories high and was the tallest building in the city at that time.

The Kansas City building presented a new challenge for Tufts, as the land was pie-shaped, which required a three-sided structure to effectively utilize the space. The final cost was $425,000 with the structure used for bottling, a distribution center, and offices. Space not occupied by Coca-Cola was leased to other tenants.

When Candler sold Coca-Cola to Ernest Woodruff in 1919, the company's local administrative offices remained in the building until 1932. However the building was sold by Woodruff in 1922.

Seven years later the building went into foreclosure and ultimately was repurchased by the Candler family, and in 1932 renamed the Candler building. In 1947 the Candlers gave the building to Emory University, and four years later it was purchased by the Western Auto Company, which had been founded in Kansas City. A large Western Auto logo sign was placed on the roof in 1952, and the company observed its 75th year in Kansas City in 1984. The unique three-sided building was placed on the National Register of Historic Places in 1988.

Advanced Auto Parts bought Western Auto in 1998, and put the building up for sale a year later. In 2004 it became "Western Auto Lofts" with 93 units offered for sale. The large Western Auto logo sign remains on the roof, but many in Kansas City still refer to the local landmark as "the Coca-Cola building".

Granite Falls, North Carolina

Looking for an old Coca-Cola vending machine for your game room? Chances are you'll find the right one in Granite Falls, North Carolina where over 2,000 restored vendors are available at Antiquities Vending Company.

Alan Huffman started his search for old vending machines in 1988, after purchasing a 1973 model at an antique store. That vendor stirred Alan's pleasant memories of purchasing cold bottles of soda from an identical machine when it was new and he was just a youngster. That 12-flavor gravity-fed soda machine sold drinks for a quarter. You could get a deposit back if you returned the bottle. Those memories sent Alan on his mission to find and restore the old vending machines.

He specializes in machines made from 1925 to the late 1970's. The most treasured find in his collection is a 1925 Coca-Cola machine said to be the first one manufactured by the ICO Company. Huffman claims it is the first vendor to be "officially approved" by the Coca-Cola Company. He also has two other old ICO machines in his collection.

Most of Huffman's restored machines are for sale. He often fabricates identical replacement parts, and boasts the restored machines "will operate and look like the day they were made."

Huffman describes his unique business as a "protective habitat for endangered soda machines". It is located at 30 S. Main Stree in Granite Falls, where over 400 models are on display in the huge main showroom. Tours are available by calling in advance at 828-962-9783.

The Candler Mansion

You may be able to spend a night at the "Candler Mansion" if plans announced in 2017 become a reality. The abandoned and long-neglected mansion near Atlanta was built in 1922 for Asa G. Candler Jr., the son of Coca-Cola founder Asa Candler.

Now known as "Briarcliff Mansion", it is part of a 42 acre site owned by Emory University. Atlanta developer Republic Property Company filed an application in 2016 to renovate the mansion and other structures on the site, including a greenhouse and carriage house. The project would be developed under a long term lease with the university.

The proposal called for the mansion to house 15 guest rooms, a full service restaurant, and event area. In addition the developer hoped to add seven new structures at the site, including guest cabins and catering facilities.

Inside the abandoned mansion

The original Candler home featured a grand ballroom, a solarium, gardens, a golf course and two swimming pools; one of which could be used by the public for a fee of 25-cents. The pool had a stand where Coca-Cola and snack items were sold, and it was illuminated at night by a neon lit fountain. The area was extensively landscaped.

In another area Candler maintained a menagerie with exotic pets including baboons, a Bengal tiger, a black leopard, four lions and six elephants. A mansion neighbor once won a $10,000 settlement from Candler, after one of his baboons jumped over a wall and devoured $60 in currency in her purse. Ultimately Candler gave all his animals to the Grant Park Zoo.

The estate was sold to the General Services Administration in 1948 for a planned veteran's hospital, which never materialized. Candler was a severe alcoholic and died in 1953. Ironically in that same year, the estate began to serve as the Georgia Clinic, the state's first alcohol treatment facility. The estate next became the Georgia Mental Health Institute, operating from 1965 to 1997 with a multi-story main building connected by underground tunnels to nearby cottages. Emory University purchased the site in 1998, and it became known as the Briarcliff Campus, but remained empty and neglected.

The new proposed development for the site was approved and gained support from the Georgia Trust for Historic Preservation and from the DeKalb History Center.

www.ingramcontent.com/pod-product-compliance
Lightning Source LLC
Chambersburg PA
CBHW050357110426
42812CB00006BA/1730